# ORGONE, REICH and EROS

## Wilhelm Reich's Theory of Life Energy

### W. EDWARD MANN

SIMON AND SCHUSTER   NEW YORK

SBN 671-21512-4
LIBRARY OF CONGRESS CATALOG CARD NUMBER: 72-86990
DESIGNED BY JACK JAGET
MANUFACTURED IN THE UNITED STATES OF AMERICA

1   2   3   4   5   6   7   8   9   10   11   12   13   14   15

The publisher wishes to thank the following for permission to reprint selections in this book. Any inadvertent omission will be corrected in future printings on notification to the publisher.

Annals of the New York Academy of Sciences, for material from an article by L. J. Ravitz, in an article, "History, Measurement and Applicability of Periodic Changes in the Electromagnetic Field in Health and Disease, 1963." With permission of the author.

Citadel Press, Inc., for material from Acupuncture and You, by Louis Moss, copyright © 1966.

Coward-McCann, Inc., for material from Pleasure, by Alexander Lowen, M.D., copyright © 1970 by Alexander Lowen.

De Vorss & Co., for material from Breakthrough to Creativity, by S. Karagulla, M.D., copyright © 1967.

Energy and Character, edited by David Boadella, for material from several issues of this journal, including articles by David Boadella and Dr. B. Bizzi.

Mark Gallert, for material from New Light on Therapeutic Energies, published by James Clark & Co., London, copyright © 1966 by M. Gallert.

Harper & Row, Publishers, Inc., for material from Explore Your Psychic World, by Ambrose A. Worrall and Olga N. Worrall, copyright © 1970 by Layman's Movement; The Gift of Healing, by Ambrose A. Worrall and Olga N. Worrall, copyright © 1965 by Harper & Row, Publishers, Inc.

Huna Research Publications, for The Secret Science at Work, by Max Freedman Long, copyright 1953; The Secret Science Behind Miracles, by Max Freedman Long.

International Journal of Sex-Economy and Orgone-Research for materials from Vol. 3, Nos. 2, 3 (October, 1944).

Interscience Work Shop, for material from Primal Scream and Genital Character, by Charles R. Kelley, copyright © 1971 by Charles R. Kelley; and material from Creative Process, Vol. 4, No. 2; and from A New Method of Weather Control, by Charles R. Kelley, copyright Interscience Research Institute, Stamford, Conn., 1961.

American Medical Association, Journal of American Medical Associa-

*Dedicated to Glendon College Library, whose atmosphere and facilities made possible many happy hours of writing.*

# CONTENTS

## PART IV

## PART V

## PART VI

# PREFACE

My first acquaintance with Wilhelm Reich and the orgone theory was in reading his *The Function of the Orgasm* in 1949. Already excited by psychoanalysis, I found Reich's extensions of Freud especially intriguing and challenging. In the early 1950's I delved more deeply into Reich's work, read around the theme of vital energies, visited him at his lab in Rangeley, Maine, and decided that the orgone theory deserved further study. Invited to a conference on healing energies at Rye, New York, in the mid-fifties, I was privileged to meet Aldous Huxley and some leading American psychics, including Mrs. Eileen Garrett, long president of the Parapsychology Foundation, and the healers Ambrose and Olga Worrall. I left that conference determined to investigate the orgone theory in depth and eventually write about it.

In the late fifties I worked at a manuscript entitled "The History of the Idea of a Cosmic Life Energy." I began in the pre-Christian era and traced the story of life-energy advocates up to the forties and Wilhelm Reich. However, I lacked the means to test rigorously the orgone theory, and at this time very little research was being carried out on Reich's work. His death in prison in 1957 had set back his "movement." Meantime, I had begun making and loaning out small three-layer orgone blankets and keeping a record of how they seemed to affect people with minor physical complaints.

Around the beginning of the seventies, interest in Reich suddenly blossomed. New pro-Reichian journals appeared, his books were reprinted, his wife's biography came out, and some professional researches were being carried out. He was being "discovered" by all sorts of people. Interest in healing energies and in ESP and in the contribution of Eastern thought to understanding life, energy and the cosmos erupted.

It was obvious the time was ripe to complete a study focused on the orgone theory and its usefulness. By good fortune, I secured a sabbatical, and the time to travel and write was available.

In the traveling and writing I was fortunate in receiving help from a number of talented people. I would like here to acknowledge my debt to Peter Davis, of Berkeley, California, to Professor David Bakan, of York University, Dr. Samuel Silverman, Dr. Leo Roy, Dr. Charles Kelley and Dr. Myron Sharaf. Two people who helped me enormously with the manuscript were Mrs. Nan Merrill, and Professor Murray Melnick, of Hofstra University. I have to thank three people who encouraged me to keep at the writing: Professor Benjamin Nelson, of the New School for Social Research; Professor Arnold Rockman, of Atkinson College; and last but far from least, my wife, Madeleine. And to my ever-faithful typist, Mrs. "Bev" Viljakainen I am also especially grateful.

I hope this study makes a contribution to what I consider a fascinating and important area of inquiry. It has been a long but devoted labor, and the very doing of it has been both memorable and instructive.

W. EDWARD MANN

*Toronto, April 1972*

# FOREWORD

By JOHN C. PIERRAKOS, M.D.    *Director,*
*The Institute for Bio-energetic Analysis*

Wilhelm Reich is a controversial figure because of the breadth
and the scope of his ideas and their impact on our times. His
philosophy swung the pendulum from the mechanism of the
scientific approach to the functional identity of man with nature and the cosmos.

On the evolutionary scale, man attempted during the primitive stages of his development to understand his environment,
within and outside himself, through his emotions and bodily
feelings, but his level of consciousness was limited. In the time
of Jesus, with chaos and destruction prevalent throughout a
world of physicality, the importance of the mind was brought
forth in order to illuminate his path of development. The
kindled flame of self-knowledge was followed, however, by
many centuries of twilight; but during the Renaissance, man
attempted again to free himself from the confusion of his emotions and feelings and the darkness of the mysticism of the
supernatural. Little by little, the Age of Reason and the Industrial Revolution came forth and gave supreme control to
the mind. However, this process divorced man from his own
physical nature, from his feelings and emotions; the pendulum
swung in the opposite direction.

Wilhelm Reich's work brings the human being back to his physical nature, and the importance it has for his life and his development. In our era, wherein scientific knowledge reigns supreme and the functions of man are further split by the exact and detailed knowledge of science and the complexity of the environment, Reich's work reverses the process of splitting and brings man back to the unity of his nature. Humanity resists this because each one of us desperately attempts to run away from the pain of the actual experience through our emotions and body. We are afraid to perceive our negativities, for fear of losing control over them, and we drive them into our unconscious, thus acting them out indirectly and in a destructive way, and by intellectualizing the basics of life. Reich recognized that the negative aspects of man as expressed in the distortion of the body and mind (his concept of *character structure*) prevent him from experiencing the flow of life within and outside himself, thus arresting his development. Reich's is a positive philosophy of life, in opposition to the basic negativity of our culture and its skepticism based on the scientific attitude of detailed objective knowledge, which so far has failed to unify the inner and outer worlds of man.

In this very extensive work which encompasses the serious efforts and labors of over fifteen years, Professor Mann takes us on a tour of the early explorations of the concepts of "life energy" to the more scientific work in American and European universities; and through parallel descriptions of these approaches he closely weaves a pattern of relationships that show that the same ideas expressed by Reich have appeared and reappeared through the history of humanity in many forms and under different names.

But first let us point out the important contributions made by Reich. Some of the most significant aspects of his work are his extensive studies of the energetic phenomena of life from the amoeba to man to the cosmos. He derived this chainlike connection of the relationships of natural phenomena by a unique process of thinking which he called *orgonomic functionalism*. The main principle of functionalism is the identity of variations in their common functioning principle. Through this approach to the realm of man, he derived the concept of psychosomatic identity—that is, whatever happens in the mind of man or animal occurs simultaneously in his body. This is different from the concept of psychosomatic parallelism in present-day medicine and psychiatry, where the physical functions are considered to occur in parallel with the emotional expressions, thus splitting their unity.

Reich came to these concepts from his extensive work in psychoanalysis, where he observed that at the beginning of therapy, there is no positive relationship with the doctor, even if the patient behaves in a positive way. This is due, he postulated, to the patient's tendency to avoid the perception of painful experiences. The patient accomplishes this by unconsciously developing physical and emotional blocks to arrest the feeling. Reich considered Freud's "libido concept" as expressing a *real energy* that flows in the organism and is regulated by laws relating to the character structure of the patient, according to an economy whereby the system permits the release or withdrawal of this energy. Reich postulated that the blocks to feeling developed early in life so as to avoid punishment or rejection of the child by his parents, and are actually muscular rigidities that regulate the flow of the feelings. He noticed that there is a relationship between the ability to flow emotionally and physically and the discharge of feeling during sex. People who are bound with blocks are unable to discharge fully, that is, to have a total orgasm, even though at times they achieve a partial release. These concepts led to the theory of the orgasm and its relationship to illness and health.

At first Reich believed that this energy, *orgone,* was specific to living organisms, but later he defined it as a universal preatomic energy. He worked with the concepts of energy and was able to develop physical methods to open up the blocks and allow the energy to flow through the body. He wanted to be known as the discoverer of this specific "preatomic universal energy," which he named *orgone,* a name derived from the words *organism* and *orgasm.* His work from the individual extended to the environment and the cosmos. He devised methods of weather control and change, and attempted to understand the functioning of the heavenly bodies, in terms of the concept of orgone energy. This is explained in Part Five of this book.

Reich's work extends to the depths of the universe, to the ramifications of the microcosm and macrocosm. He also contributed significantly to many fields of knowledge, such as psychology, sociology, medicine, biology, pathology, agriculture and meteorology. In our civilization there are few beings who, through their understanding of the life processes, become the navigators of life. Wilhelm Reich has navigated humanity to the depths of its biological existence.

However, many difficulties that prevented the acceptance of his work, besides man's natural resistance to revolutionary views, may be related to the nature of his personality. Reich, like any other human being, had his flaws, but this should not detract from the importance and the greatness of his work. According to many of his co-workers, Reich was very human, natural and childlike. He was a natural scientist. All his findings were reported, but there was a constant feeling of urgency on his part about his work being accepted and communicated to important people. In my contact with him, I perceived him as a soft and warm being on one side, but on the other side, very quick to anger, and violent in his reactions. He reacted under adverse conditions with sadness, hurt and the feeling that he was a misunderstood genius. He felt lonely and isolated, but inside he was soft and yielding. The outer surface presented at times a picture of a hard shell and arrogance.

Reich showed a lack of sentimentality. He despised sentimentality. In his dealings with other people, he sometimes acted as though he was separate and above, and thus he alienated important people interested in his work. He was very proud and was living up to his image, which was that of an important discoverer of the phenomena of life. There was a personal loneliness and an apparent separation from humanity in his relationships, in spite of his understanding and extensive descriptions of the psychology of people. In his dealings with women, according to the biography by his former wife, Ilse Reich, he was gentle and loving and let himself be open. He seemed driven and obsessed with his work. He had a deep cosmic sense, but I believe that he could not make the final connections with the soul concept as Jung did. He never accepted true spirituality in his published works.

But let us follow this further and make some historical connections in terms of his background and his cultural milieu.
Around the end of the nineteenth century, there was a violent
counterreaction against energy concepts as propounded on
the Continent in the work of Mesmer, Reichenbach and others.
In England this work was taken up and connected with the
spiritualistic phenomena which important scientists like Professor William Gregory, Sir William Crookes and Sir Oliver
Lodge had undertaken to investigate seriously. On the Continent, however, it took the form of experimentation with
hypnosis. Let us keep in mind that Freud discovered the forces
of the unconscious, having worked in the clinic of Liébault at
Nantes with hypnotic and trance phenomena, and subsequently
with Breuer on the energetic phenomena of hysteria. Freud,
even though he developed the concept of libido as a real energetic concept, later rejected both the energetic aspect of libido
and the spirituality forming the center of Jung's work.

At the time of Reich's upbringing, there was great emphasis
on the importance of the mind and the unconscious, due to
Freud's work. The spirituality of his time in Austria and Germany was represented by low-level spiritism which was considered to be rank superstition. In view of the condemnation
of the work of Mesmer and his followers, and the British
spiritualists, Reich did not see the other side of true spirituality
to which Jung devoted his whole life. It appears that Reich
must have had a serious problem there which he never worked
out. He called it mysticism and explained it as a split in the
unity of the structure. He fell into the common mistake of rejecting all spirituality because of the excesses and charlatanism
to which it too frequently lends itself. He did not differentiate
between spirituality and distorted mysticism, the latter being
an escapist and split function of man. In several later books
he idealized his self-image and presented himself as a savior
of the world, a great messiah of the modern times; his was the
only truth. The orgone energy became synonymous with God,
a supreme force that explained all the phenomena of life. Such
concepts led inadvertently to a cultish attitude by some of his
followers.

Looking with perspective at this great era, in which the discovery of the unconscious was made, it has to be pointed out

that while Freud worked primarily with the mind of man, Jung focused his work on the relationship of the soul to man as an entity, and Reich focused on the biological and physical expressions of man as related to the whole being. The three men represent the three aspects of man: the body, the mind and the spirit.

Reich was very disappointed at the end, when he felt that his labors and efforts and the image he had of himself were not fulfilled. One wonders whether there was an unconscious intent in his martyrdom, since by handling his affairs with the Food and Drug Administration in the way he did, he died in prison from heart disease one day before his release. Is it possible to be a martyr without unconsciously wanting it?

Having described in a general way some of the essential issues of the life, tenets and personality of Reich, I would like to go specifically into the subject matter of this book, which Professor Mann assembled with great effort over a long period of time, consulting with many people, reading extensively, and performing himself some of the experiments with accumulator devices.

The book is divided into six parts. Part I begins with a detailed and extensive description of the discovery of orgone energy and the concepts of this energy in the body. This is a factual presentation of the work taken from the original articles and quoted very accurately. Part II deals primarily with historical aspects of the exploration of the life energy in different cultures and civilizations; it includes the Egyptians, the Chinese, Biblical times, the work of the great healers of the Renaissance, Paracelsus, and Jan Baptista van Helmont, the Belgian physician. All of these confirm the existence of a life energy. In these sections of the book, Professor Mann presents the work of some of the great healers, such as Valentine Graterakes, and the work of scientists like Luigi Galvani, who opposed the mechanical interpretation of energy phenomena of Volta and Fabrizi. He then proceeds to describe the work of Mesmer and von Reichenbach, and to compare it with the concept of orgone energy as expounded by Reichians and neo-Reichians. At the end of the first chapter in Part II there is a comparative summary of the concepts of Mesmer and Reich which reveals an amazing similarity, though the two sets of concepts were arrived at independently. Also presented is the idea of an electrodynamic force field, which was developed mainly at Yale University by Burr, and at William and Mary College by Ravitz. This work too shows a similarity to the energetic concept of the orgone. Then in Part III we see how Indian philosophies, the auric energies and the different concepts of energy as applied by yogis and magnetic healers relate to the orgone. This is done with love and care, bringing together a great amount of information relevant to the subject matter. In Part IV Professor Mann presents various applications of the orgone energy. The nature of the cancer problem is very well explained. The work of some of Reich's collaborators and of independent researchers like Dr. Charles Kelley and Dr. Bernard Grad is outlined, as well as some of my own observations of the energy field. This section also attempts to present further applications of the concept of the orgone into the atmosphere and the relationship it has to weather formation; a detailed account is given of weather making. Some of the work of the Italians interested in Reich is mentioned, and a documentation of their results is presented.

In Part V accurate and detailed information is given about the latest work in the borderline sciences that use energy concepts similar to the orgone. These are the fields of bioelectricity, biomagnetism, biotronics. Under the last heading, the work of the Russian and Czech experimenters is brought into focus as they define the specific biological energy as bioplasma. A special chapter is devoted to the work of the Russian experimenters, in which the Kirlian photographic phenomena of the auric field are explained. Findings of various researchers as expressed in the concepts of biological clocks and the cyclical rhythms of events in relation to astrology are presented.

Finally, in Part VI, an attempt is made to connect Reich's work with various philosophies of the European and Eastern scene and hammer out his basic contributions in such fields of endeavor as psychoanalysis, therapy, weather control, and the specific methodology that Reich used in connecting the body and the mind through the concepts of psychosomatic unity. Also, a preliminary explanation is given why people who are innovators meet with such great resistance from their respective cultures and why they are crucified sooner or later. In this part, Professor Mann deals only with the objective and external factors creating these reactions.

It is of great interest that the therapeutic methods developed by Reich through orgonomy, the science of orgone energy, and the development and broadening of these techniques in the work of bioenergetic analysis as developed by Dr. Alexander Lowen and myself, are today in the center of the human-potential movement and of group-encounter and gestalt therapies. These are the new therapies which came mainly from the people and not from the psychoanalytic movement as such. They were formulated by farsighted psychologists, scientists, educators, artists and "feeling individuals," and are practiced today in hundreds of clinics, private offices and growth centers throughout the United States, Canada, Europe and South America. In these therapies, there is an attempt to reach a deeper level of communication among human beings, allowing the negative feelings to be recognized, to be understood by encountering them in each other's reactions and attempting to integrate them into mature-adult behavior. This very important movement of today has incorporated the knowledge of body processes as a focal point; the character-analysis work of Reich and our own work in bioenergetics form its physical anchor.

This movement reflects the swing of the pendulum away from scientific materialism and into spirituality and the experiential. Of course, there are excesses harmful to young people through abusing the intake of drugs. However, the physical work of bioenergetics which helps to establish the identity of man with his body can bring about a valuable balance. The deeper a person feeds his physical roots, the more branches, flowers and fruit he produces, the higher the spirit soars into the cosmos. This is the era of Aquarius, of group enlightenment, and of deep and genuine spirituality.

Finally, the book is a testing ground for a variety of ideas intended to bring a new approach to the thinking of man, away from mechanical scientism and into a way which incorporates the subjective and objective approach to knowledge. I feel that this work will contribute immensely to opening new roads to many scientists who are fixated on theories derived from the Newtonian concepts of definite mass and energy. It is a borderline area that has to be explored by man in order to reach a sound and unified concept of life based on the greater realities of the cosmos, the physical, the mental and the spiritual. To paraphrase Shakespeare, "there are many things between heaven and earth not known to man."

# Introduction

Since the famous formula of Einstein and the production of nuclear energy, the twentieth century has emerged as the century obsessed by energy. The day-to-day operation of a technological civilization demands this preoccupation, as does man's restless, energy-draining search to conquer more of nature. It was both logical and less anxiety-creating that the early inquiries were focused on energy in the universe and in matter, and only much later began to accept living organisms and man as energy systems. After the First World War, philosophy in the person of Bergson elaborated a vitalistic perspective, but not till recently have many scientists concentrated on the energies that lie within and drive organisms. The elucidation and application of such vitalities may lead to a "bomb" as potent in its own way as that exploded at Hiroshima.

One man who became obsessed by the question of vital energies in organisms, nature and man was the German-born psychoanalyst colleague of Sigmund Freud, Dr. Wilhelm Reich. During his lifetime—he died in 1957—his numerous books and articles on what he called orgone energy excited both interest and considerable persecution. Now, some fifteen years later, one finds new interest in him and his orgone theory. Thus, a review of his *Mass Psychology of Fascism*, in *The New York Times* (January 4, 1971), has the reviewer Lehmann-

Haupt saying: "Wilhelm Reich has made a comeback." He
notes that "Mailer called Reich 'hip' as a mind but 'square' as
a stylist" and adds that the book "surpasses Marcuse's attempt to reconcile Freud and Marx." And he concludes that
"whereas fifteen years ago this reviewer contemplated Reich's
theories of sex economy and orgone energy with horrified
shudders, reading *Mass Psychology of Fascism* today makes
him wonder a little . . . is it time to reopen the question of
cosmic orgone energy?"*

While similar articles appeared in British intellectual journals in 1970,† *The New York Times* went all out on April 18,
1971, by devoting a 6,000-word article to Reich and the orgone. Written by Professor D. Elkind, the article took a critical stance near the end, but was generally not too unfavorable.
In short, in the world of intellectual sophistication, Reich is
becoming a name to conjure with.

The difficulties of grappling with the orgone-energy theory
are partly illuminated by a simple sketch of Reich's professional career. In 1919 he attached himself—at the age of
twenty-two, and while still attending medical school—to that
way-out, exotic group of agnostic Jews who, clustering around
Sigmund Freud in Vienna, laid the groundwork for the profession of psychoanalysis. Among their contemporaries, Freud
and his eager disciples were regarded as kooks, oddballs,
freaks, out to sexualize the world. How Freud won acceptance
for his unorthodox ideas, and how psychoanalysis—which
began in Austria and Germany—became a "big thing" in the
United States is in itself a fascinating story, a story commented on by biographers and in-group members, but still
lacking definitive sociological treatment.

Beginning as an enthusiastic follower of Freud, Wilhelm
Reich became a devoted colleague and was stimulated to
launch some distinguished and independent work in the late
twenties. A vigorous member of the International Psychoanalytic Association and a practicing analyst, he wrote a
number of well-received articles and then several books before
he turned thirty. Within this fringe group of alienated intel-

---

* *The New York Times*, January 4, 1971, Section VI, p. 29.
† "The Return of Reich" and "Liberation by Orgasm," in *New Society*,
September 3, 1970, and March 28, 1968.

lectuals, he became the leader of those committed to (1) integrating Marx and Freud, and (2) combining theory and clinical work with practical preventive efforts to erase human neuroses. As early as 1929, for example, he joined the Communist party and began to operate sex-hygiene clinics in Berlin and elsewhere, designed to free proletarian youth from puritan and authoritarian hangups. (These might be called sex information clinics today.) Arthur Koestler, in *The God That Failed*, commented on his contact with Reich in a Berlin Communist cell; he said: "Only through a full uninhibited release of the sexual urge could the working class achieve its revolutionary potentialities and historic mission; the whole thing was less cockeyed than it sounds."

By 1930 Reich had become too independent for Freud, and they began to grow apart. Reich's emphasis on the importance of negative transference (and its handling) for psychoanalytic therapy, and on character armoring and orgastic potency as the test for therapeutic success was putting a serious strain on Freud's friendship. By 1934 he had proved too deviant for both the German Communist party and the International Psychoanalytic Association, and each in its own way unceremoniously kicked him out. His criticisms of "red Fascism" in Russia and of Freud's death-wish theory were prominent reasons for his rejection by these two movements.

By 1935, Wilhelm Reich, now a refugee from Nazi Germany, wandered around Europe anxiously looking for a professional home and a new and secure perspective. He determined to discover the physical character of the libido energy that Freud early postulated. In Norway he engaged in lab experiments measuring the electrical components of various forms of sexual interaction. As early as 1934 he wrote an article entitled "The Orgasm as an Electrophysiological Discharge" (*Journal of Orgonomy*, Vol. 2, No. 2 [Nov. 1968], p. 717). Subsequent investigation led to the discovery of what he called "bions"—basic life-energy units—and the origin of a new field of study, biophysics and orgone (life) energy. By 1938, these researches had aroused the ire of the Norwegian press, and he was forced to sail for America, which became for him "The Promised Land." Rejected by the bulk of European psychoanalysts, he became marginal to that highly marginal medical group, psychoanalysis, and ended up in New York a marginal man in the land of liberty, heretics and outcasts.

Initially he lectured at the radical and, at that time, not widely recognized New School for Social Research, in New York. Then he began to build a lab in Forest Hills, New York, and a new form of therapy called "vegetotherapy." This is now practiced as psychiatric orgone therapy by Dr. Ellsworth Baker in New York and by others in the College of Orgonomy. Alexander Lowen's bioenergetics is a modification of this. Meanwhile, he developed the concept of orgone energy and began experiments to test orgone's effect on cancer, using various kinds of orgone-energy accumulators. Efforts to get Albert Einstein to validate the existence of this new energy, launched in December, 1940, ended in dismal failure within a few months.

Around 1945 Wilhelm Reich added to his isolation and ultimately to his eccentricity by leaving New York to settle on farm property near the tiny town of Rangeley, Maine. There he built a small laboratory, gathered around himself a few score followers—mostly unconventional doctors, psychiatrists and oddballs—and delved further into the secrets of his orgone energy.

In 1947, partly as a result of two inaccurate articles by Mildred Brady, a free-lance journalist writing in the *New Republic* and *Harper's*,* Wilhelm Reich was suddenly labeled dangerous and a schizophrenic. Behind this attack were some uneasy psychiatrists and psychoanalysts. Even the Menninger Clinic leaders accepted the Brady story as reliable. Soon, the United States government's Food and Drug Administration began an investigation of his orgone accumulator, which he now claimed to be useful in cancer therapy but not a cure. As his permissive views on sex for teen-agers, written in the early thirties, leaked out, circulation-hungry magazines put out sensational, often inaccurate and usually "put-down" articles. For instance, one published by *Sir* Magazine, November, 1951, dealt with "California's Strong Love Cult."

Somewhere around 1950, Reich became labeled among the United States medical fraternity as a quack. The orgone ac-

---

* *New Republic*, May 26, 1947; *Harper's*, April, 1947.

cumulators, rented out for ten dollars a month to, at most, several hundred users, mainly doctors and therapists, were supposedly bringing in a lot of money and doing nothing for the patients. (The income was used for research expenses.) Now, completely outside regular medicine as well as psychiatry and psychoanalysis, Reich, with his own journal, printing house and band of followers called medical orgonomists, had become totally marginal. In 1949 he tried to test the power of the orgone to control atomic energy, but the resultant intermixture produced physically dangerous effects to his band of experimenters, and some of his followers quit in anger or fear.

During the fifties things went from bad to worse. In 1954, the Food and Drug Administration closed in, suing him for mislabeling the accumulator as a cancer cure. He fought the charge fiercely, claiming the government had no right to persecute him for what were purely scientific activities. Rather than defend himself against the specific charges of the government, he lashed back at them, antagonized and then lost his own lawyer and finally refused to enter the court for the trial. He was finally convicted on a contempt-of-court charge and was sentenced to two years in a federal prison. Shortly after, one of his close associates killed himself and some of his close followers began to wonder about his mental health. The English radical schoolmaster, A. S. Neill, of Summerhill, a close friend to the end, admitted he could not follow Reich into certain latter-day applications of the orgone theory. His circle of friends and followers diminished. This meant that during the years when he expanded the orgone theory to include meteorological and cosmic forces and even flying saucers, the circle of potentially friendly critics dwindled. Those left were mainly too conscious of his greatness to be critical of grandiose theory building. Then, when he was put in prison, the government also burned or tried to confiscate all his books that mentioned the orgone. When, nine months later, he died in prison, it seemed the end of a once brilliant but—to many persons— very unstable eccentric.

For twenty-five years the widely accepted view of Wilhelm Reich was that he had written a masterful and original book in *Character Analysis* and that other early work showed brilliance but that with the orgone-energy theory he had "gone off his rocker." The idea of a cosmic life energy as the essential element in all biological life was dismissed as the fond creation of a good mind overwrought with the need to physicalize Freud's abstract concept of the libido. On every side, medical, psychiatric or literary, this cornerstone of Reich's creative work was ridiculed: for years he was called mad, a crackpot, a crank or a quack. Thus, in his *Developments in Psychoanalysis*, the psychiatrist Leon Salzman, who credits Reich with some useful contributions to psychoanalysis, says: "Reich was either a genius or a madman depending on one's point of view . . . in the pre-orgone phase Reich was a brilliant traditional analyst" (p. 19). Beka Doherty in a book, *Cancer*, published in 1949, said: "The orgone accumulator is based on ideas which have just enough scientific possibility to captivate people" (p. 53). A typical dismissal of the orgone and its utility with cancer was presented in a letter to *The New York Times Magazine*, May 16, 1971, by a Dr. Brussel, Assistant Commissioner of the New York State Department of Mental Hygiene, 1952–69, who ended up a damning attack on several Reichian psychiatrists by saying, "Reich's phony cancer cure still stands as one of the most dastardly and dangerous quack therapies ever perpetrated by a man who once enjoyed the respect of the medical profession." *Time* magazine, in a major piece on the sex revolution, on January 24, 1964, gave expression to conventional oversimplifications and distortions about Reich and the orgone in the following words, which opened the article:

> The Orgone Box is a half-forgotten invention of the late Dr. Wilhelm Reich, one of Sigmund Freud's more brilliant disciples, who in his middle years turned into an almost classic specimen of the mad scientist. The device was supposed to gather . . . that life force which Freud called libido . . . The narrow box, simply constructed of wood and lined with sheet metal, offered cures for almost all the ills of civilization and of the body; it was also widely believed to act, for the person sitting inside it, as a powerful sex stimulant. . . .

Latterly, not everyone has been so certain about the orgone or its accumulator. For one thing, Reich has been proved by events to be a brilliant social prophet. For instance, after a trip to Russia in the late twenties, he labeled its society as "red Fascism." His book, published shortly afterward, entitled *The Sexual Revolution*, forecast much of the present thinking and behavior of present-day youth on sexual issues. "Reich [also] had argued that the dissolution of the traditional middle-class family would lead to the disappearance of the Oedipus complex [and the experiences of the Israeli *kibbutzim* would later prove him right]" (P. Roazen, *Saturday Review*, February 13, 1971, p. 49).

Again, writing in the very early fifties in *Cosmic Superimposition*, he claimed the earth is enveloped in a blue dynamic-energy field, something only recently established by the experiences and photography of the astronauts. In addition to these and other "predictions," Reich's uninhibited critique of mechanistic, antilife science forecast the outcries of radical segments in the educational and scientific communities as they protest science's genuflections to militarism, money and authoritarianism. Thus, a few respected American and European writers now are considering Reich as a significant social prophet—and wonder whether he might not have been at least partly right about a life energy.

A central issue is that so far, no widely respected scientist or scientific group has bothered to try to test the basic experiments upon which Reich erected his theory. So his followers claim, in the absence of negative evidence, science should at least look at the theory—and then give it a fair test.

Actually the question of whether the orgone had been scientifically tested may be characterized as a semantic issue. How much testing is required and by whom, and with what results? Which group is calling the shots—that is, large research institutes, formal academicians or friends and followers of the man? The Food and Drug Administration claimed during the court hearings that they had commissioned independent tests and that those tests indicated there was no such entity as orgone energy. When questioned as to the character of the testing and the scientists involved, they refuse to say anything. Various interested persons, including myself, have received the same answer. (A photostat copy of their letter to me is included in the Appendix.) Friends of Reich claim that in the court hearings, Food and Drug Administration scientists admit they found *some* evidence for the orgone.* They say the mice placed in the accumulator died more rapidly than the controls, the reason being—according to Reichians—that it was placed near operative X-ray machines. (Theoretically, X-ray radiation overexcites the orgone and makes it destructive.) They also point out that as early as 1941, Einstein after a five-hour chat with Reich, went over in Reich's absence a basic experiment showing orgone's temperature effect (a temperature difference of 1–2°C.) and admitted he found the effect, but claimed it was explainable in terms of heat convection. This will be described in detail in Chapter One.

Scientists favorable to Reich wrote articles in his journal during the late forties, giving positive results in the use of the accumulator. Other periodicals would not accept such articles. A competent psychologist researcher, with an entry in *American Men of Science*, replicated various tests of the orgone in the early sixties and published them in his journal, *The Creative Process*. But this was given no attention. This scientist's requests for funds from various research groups and foundations were turned down repeatedly in the next few years. In the early fifties I made a serious attempt to interest the chief officers of the Canadian Cancer Association in tests on cancer or arthritic patients, but the response was totally negative.

Official science has had no time for, or interest in, trying to corroborate Reich's theory of the orgone. A similar thing happened in 1937 in Norway: medical scientists claimed his technique with the microscope was defective, and they refused to organize a scientific group to actually work with Reich to test his theory of the bion.

Recently, however, a variety of scientific investigators working independently in fields like biology, medicine, parapsychology and biophysics, have unearthed evidence of energies with characteristics that look startlingly like those claimed for the orgone. In America these include former Yale biologist Harold S. Burr and L. J. Ravitz and associates, of William and Mary universities, who have apparently established the exist-

---

* A critical analysis of these tests appears in the *Journal of Orgonomy* (Box 476, Ansonia Station, New York, N.Y. 10023), Fall, 1972.

ence of energy force fields around plant, animal and human organisms; the investigation into psychic energies displayed by clairvoyants and sensitives by a highly regarded psychiatrist and onetime collaborator of Dr. William Penfield, Dr. Shafica Karagulla of the Institute for Higher Sense Perception; and the newly begun investigations of the Life Energies Research Foundation of New York, headed by the well-regarded psychiatrist, Dr. Robert Laidlaw, a long-time friend of Mrs. Eileen Garrett, organizer of the Parapsychology Foundation. These and other equally fascinating investigations will be outlined in detail.

At this point, an outline of the orgone theory may be helpful. As Reich conceived it, orgone energy is present in the atmosphere, is related to the sun, extends through all space like the "ether," is drawn in by all organisms, and is what accounts for the movement—contraction and expansion—of all living things. It flows through organisms, creates a field around them and can be transmitted from organism to organism (among human beings, by the laying on of hands, for instance). It governs the total organism and expresses itself in the emotions as well as in purely biophysical movements. In the sexual orgasm a large discharge of orgone takes place, whose biological function is to restore energy equilibrium to the organism. If the orgone flow is unnaturally checked in an organism (e.g., by character armoring), disease will set in. It is believed that orgone has a strong affinity for water. It holds together the moisture elements in clouds. It is the basic link between inorganic and organic matter.

The following ten-point statement, a quick summary of the theory, was produced by Dr. Charles R. Kelley, editor of the journal *The Creative Process:**

1.  Being mass free, orgone energy itself has no inertia or weight. This, it is noted, is one of the main reasons why it is difficult to measure with conventional techniques.
2.  Present everywhere it fills all space, although in differing degrees or concentrations. It is even present in vacuums.

---

\* This list of properties of the orgone was first summarized in C. R. Kelley, "What Is Orgone Energy," *The Creative Process*, Vol. II, No. 1, pp. 58–80. It is presented here with slight changes, with permission.

3.  It is the medium for electromagnetic and gravitational ac-
    tivity. It is held to be the substratum of the most funda-
    mental natural phenomena, the medium in which light
    moves and electromagnetic and gravitational fields exert
    force.

4.  Orgone energy is in constant motion, and this can be ob-
    served under appropriate conditions. For instance, the
    bluish heat waves seen shimmering above wooded areas
    and mountains are said to be orgone energy movements.
    Its motion has at least two characteristics, a pulsation
    form—that is, alternating expansion and contraction—and
    a flow normally along a curving path.

5.  It "contradicts" the law of entropy. Orgone energy is at-
    tracted to concentrations of itself. Unlike heat or elec-
    tricity, which manifests a direction of flow from higher to
    lower potential, orgone flows from low orgonotic poten-
    tials to higher. Thus, high concentrations of orgone
    energy attract orgone from their less orgone-concentrated
    surroundings. (An analogy is found in gravitation, where
    the larger bodies attract, or "pull," the smaller.) Non-
    entropic orgonotic processes, moreover, do not run their
    course mechanically, but are qualitatively entirely differ-
    ent from entropic processes.

6.  *It forms units or entities which are the foci of creative
    activity.* These orgone energy units may be living or non-
    living. The living ones include bions, cells, plants and
    animals, and the nonliving include clouds, storms, planets,
    stars and galaxies. All of these orgone energy units have
    *certain* features in common. For instance, all are "nega-
    tively entropic" in the sense mentioned above, so that
    they acquire energy from their environment and all have
    a "life cycle" passing through birth, growth, maturity and
    decline.

7.  Matter is created from orgone energy. Under appropriate
    conditions matter arises from mass-free orgone. These
    conditions are held to be neither rare nor unusual.

8.  It is responsible for life. Orgone energy is the *life* energy
    and as such is responsible for the special characteristics
    which differentiate the living from the nonliving.

9.  Separate streams of orgone energy may be attracted to

each other and then superimpose. The superimposition function is held to be the fundamental form of the creative process. Thus, in free space, superimposing orgone-energy streams typically show the form of two streams of energy converging in a spiral. This is seen in spiral galaxies and also in the structure of hurricanes and other cyclonic storms. Celestial functions such as sunspot cycles, aurora borealis, hurricanes, tides and major weather phenomena are considered expressions of the interplay of two or more cosmic energy systems and also involve spiral forms of superimposition. In living nature, mating is a principal expression of superimposition: two separate streams of energy flow together and superimpose during the coital act.

10. It can be manipulated and controlled by orgone energy devices, the best-known being the accumulator . . . Certain experiments indicate that the air temperature within the accumulator, and the body temperature of anyone sitting in it, rises, up to one degree centigrade, with variations depending upon the outside weather, the time of day and the sitter's character structure. Other evidence includes an increase in the impulse rate of the Geiger-Müller counter when exposed to orgone concentrations in an accumulator.

*Human beings are deep wells of energy, immense reservoirs of force, the force of life determined to subdue matter.*
—COLIN WILSON

# PART
# I

*Chapter One*

# REICH'S DISCOVERY OF ORGONE ENERGY

Wilhelm Reich was a man of tremendous physical energy, with an easy acceptance and expression of sexuality. A strong interest in biology and health was natural to him, for he began life on a farm and even ran the farm himself for a time in his teens. His interest in energy theories was evident early. Before receiving his M.D. he wrote an article entitled "The Concept of Psychic Drives from Forel to Jung." This was followed, shortly after he got his degree in 1922, by another article, "The Energetics of Drives." As he trained under Freud and as he practiced psychoanalysis, he soon became convinced that of all the drives, the sexual was the most potent.

While not denying the importance of the sex drive for the procreation of life, Reich early zeroed in on the energy functions of the orgasm. He finally came to view the orgasm as holding the key to the regulation of bioenergy, for by it, he felt, is effected the discharge of that part of life energy that is not consumed in the other activities of the human being. In addition, he saw it furnishing the well-being and the pleasure that make life enjoyable. In order to fully discharge the organism's "surplus" energy, Reich concluded in his book *The Function of the Orgasm*, a certain kind of "healthy" orgasm was essential. From clinical experience he believed that inadequate orgasms left surplus energy in the body that could power sec-

ondary and often unhealthy drives. Raknes sums up Reich's
insights on orgasm as follows: "A healthy sex life depends on
a complete convulsive discharge of sexual energy in the embrace of a beloved partner . . . with momentary loss of consciousness. The capacity for such an experience Reich termed
orgastic potency" (Raknes, 23). Continued work led Reich to
postulate an orgasm formula—later called the formula of
biological tension and discharge. This formula pinpointed the
energy process in the orgasm, as having four beats: mechanical
tension → bioelectric charge → bioelectric discharge → relaxation.

In the early thirties, Reich's researches on the electrical
components of erotic acts and the orgasm led to a new set of
inquiries, which by 1939 revealed the existence of the orgone.
Experiments conducted at the Institute of Psychology of the
University of Oslo (from 1934 to 1936) had measured on an
oscillograph the variation in the skin's electrical potential when
it was touched in a way that provoked pleasure or anxiety.
Results showed that skin potential climbed with pleasurable
feeling and fell with anxiety. Reich became curious to see if
his orgasm formula and reflex could be found in other parts of
nature. So he examined the movements of very small units of
life under the microscope. Out of this new level of investigation he came upon analogies between the orgasm formula and
the movements first of small moss vesicles and then of other
small vesicles observed microscopically.

By way of explanation Reich stated:

> The orgasm is a fundamental biological phenomenon; funda-
> mental because the orgastic energy discharge takes place in the
> form of an involuntary contraction and expansion of the total
> plasma system . . . Biophysically speaking, it is impossible to
> distinguish the total contraction of an amoeba from the orgas-
> tic contraction of a multicellular organism. The outstanding
> phenomena are intensive biological excitation, repeated expan-
> sion and contraction, ejaculation of body fluids in the contrac-
> tion, and rapid reduction of the biological excitation . . . In
> its quickly alternating expansions and contractions, the orgasm
> shows a function which is composed of tension and relaxation,
> charge and discharge: *biological pulsation.*

Closer investigation reveals the fact that these four functions appear in a definite four-beat: the mechanical tension, which shows itself as sexual excitation, is followed by a bioenergetic charge of the periphery of the organism. This fact was unequivocally demonstrated by measuring the bio-electric potentials occurring with pleasurable excitation of the erogenous zones. When tension and charge have reached a certain degree, there occur contractions of the total biological system. The high peripheral charge of the organism is discharged. This is seen objectively in the rapid decrease of the bio-electric skin potential; it is felt subjectively as a rapid decrease in excitation. . . . We know from earlier investigations that the function of tension and charge is characteristic not only of the orgasm. It applies to all functions of the autonomic life system. The heart, the intestines, the urinary bladder, the lungs, all function according to this rhythm. Cell division also follows this four-beat. So does the movement of protozoa as well as metazoa. . . . This biological basic formula comprehends the essence of living functioning. The orgasm formula shows itself to be the [basic] life formula as such. [Gallert, pp. 61–62]

From this, Reich went on to conclude that every living organism is a membranous structure that contains an amount of orgone in its body fluids; it is an "orgonotic" system. Thus, the term *orgone* comes from "organism" and "orgasm," and means an energy found within all organisms and basic to the orgasm reflex.

Researches in Norway late in the thirties led Reich to postulate the existence of *bions*—that is, basic units of orgone. These he produced from many different materials as reported in the book *Die Bions* (Oslo: Sexpol Verlag, 1968). In these experiments Reich took organic material like dry moss or grass, heated it to a high temperature and then let it swell in sterilized water. Microscopic examination showed that small vesicles would detach themselves, that they would expand and contract like protozoa. After a while, they would gather in heaps, surround themselves with membranes and start moving like protozoa. In other experiments, inorganic materials like coal and dust or rust were heated to incandescence in a gas flame and, while aglow, were put into a sterile nutritive solution. While in this solution, some of the particles would swell into vesicles that moved like the others of more organic origin. Although critics thought these vesicle movements were well-known molecular Brownian movements, Reich and his associates denied this and claimed they were soft and organic and showed inner pulsation, while the latter were angular and mechanical. Reich labeled these energy vesicles *bions;* and in their tendency to organize into cells like protozoa, he saw a demonstration of the very origin of life.

Certain bions were produced from beach sand, and the resulting growth, when inoculated on egg medium and agar, consisted of large, intensely blue packets of vesicles.

> Examination under a microscope with magnification of 2–4,000x showed forms which refracted light strongly, consisting of packets of six to ten vesicles and measuring about ten to fifteen microns. These were termed *SAPA* bions [from the words *sand* and *packet*]. They showed characteristics of unusual interest.
>
> When brought near cancer cells, the SAPA bions killed or paralyzed those cells even at a distance of ten microns. When cancer cells came as close as that to these bions, they remained in one spot, as if paralyzed. They would turn around and around in the same spot and finally become immobile. Dr. Reich recorded this phenomenon on microfilm.
>
> It was found that daily observation of the SAPA bions through the microscope resulted in eye irritations and conjunctivitis of the eyelids, to a far greater extent than occurred from the same duration of microscope use with other objects. This was the first effect noted, of what was later termed "orgone radiation." Attempts to identify the nature of this radiation and discover its characteristics, at first met with many failures . . . It was then found that SAPA bion cultures, placed on a glass slide held in the palm of the hand, produced discoloration of the skin, and after repeated application, an inflamed and painful condition of the palm.
>
> The air in the room where the cultures were kept became extremely "heavy," and people who stayed in the room developed headaches if the windows were closed for as short a time as one hour. [Gallert, p. 63]

This convinced Reich that he was dealing with a radiation. Later, he observed that metal objects like scissors and needles kept nearby had become magnetized. (This observation was not fully followed up.) Attempts to capture the SAPA radiation on various kinds of photographic plates led to the discovery that all got fogged *including the control plates not exposed to it*, but kept in the same room. It seemed that the radiation was everywhere!

"Peculiar optical phenomena developed when dozens of the SAPA bion cultures were prepared and kept in the room simultaneously" (Gallert, pp. 62–63). Reich saw foglike emanations and bluish dots and lines. A kind of violet light

> seemed to emanate from the walls as well as from various objects in the room. A magnifying glass held before the eyes made these impressions more intense, and the individual lines and dots became larger, showing that they had an objective reality . . . owing to the intensified amount of this energy in the dark room, the eyes of an observer would become red and painful after just an hour or two, even when no microscope was used. More prolonged exposure resulted in a curiously luminous phenomenon in which one could see a radiation from one's palm, shirt sleeves and, with a mirror, from one's hair. The blue glimmer would then be visible as a slowly-moving, grey-blue vapour around one's body and around objects in the room. People who were told nothing about these effects, observed exactly the same phenomena when placed in the room in the dark. [Gallert, p. 63]

It seems that this concentration of SAPA bions made human auras and rays from the hands visible to the nonclairvoyant observer.

Later Reich tested the SAPA bions with the electroscope, which reacts to static electricity and ionizing radiations. At first the reactions were nil. But, when rubber gloves that had been left near the SAPA cultures—and had apparently absorbed their radiant energy—were placed near the electroscope, they produced a strong reaction.

> It was then found that paper, cotton, cellulose and other *organic* insulating substances took up an energy from the SAPA

bion cultures which gave a reaction in the electroscope. High humidity, ventilation in the shade, or touching the substances with the hands for several minutes, all eliminated the effect. The fact that the radiation from the cultures would affect an electroscope only through the indirect route of being used first to "charge" an insulating material, was evidence to Reich that the radiation was of a hitherto unknown type, with different characteristics from any then known to science. [Gallert, p. 64]

Ordinary rubber gloves exposed to sunlight or to UV light received the same charging effect. Placing them on the stomach of a vegetatively active person for five to fifteen minutes charged them similarly. With biophysically weak persons, the electroscopic reaction (to the gloves) was slighter. "Increased respiration made it stronger" (Reich, *Cancer Biopathy*, p. 77). Reich concluded after months of repeated experiments that the energy was present in organisms, which take it from the atmosphere, and that some of it is everywhere.

The next step in the study of the orgone was to build an enclosed space within which the SAPA bions could be studied. As Reich states it:

No organic material could be used for this purpose, since, as we have seen, organic material absorbs this energy. According to my observations, metal, on the other hand, would reflect the energy and confine it within the enclosed space. However, the metal would reflect the energy also to the outside. In order to avoid this, the apparatus had to have metal walls on the inside, and walls of organic material on the outside. With this construction, it was to be expected that the radiations from the cultures would be reflected by the inner metal walls, while the outer layer of organic material (cotton or wood) would prevent or at least reduce the reflections to the outside. The front wall of the apparatus was to have an opening, with a lens through which the energy could be observed from the outside.

In order to assure the visibility of the orgone rays, it was decided to use a disc of organic material in conjunction with a lens system. A cellulose disc in a film-type viewer was used . . . It was found that the radiations from SAPA bion cultures placed in the box gave bluish moving vapours and light-yellowish points and lines. But the experimenters were puzzled to note the same visual phenomena in the box after it had been ventilated; then in the box after the parts had been taken apart, aired, and the metal plates dipped in water to eliminate all traces of orgone energy; and finally, in a completely new box built for the purpose, which was kept away from the SAPA cultures.

These experiments served to emphasize the deduction that had first been made from the work with the electroscope, that the energy came from the sun and also was present in the atmosphere everywhere. [Gallert, pp. 64–65]

In brief, the boxes which began as observation chambers were found to "draw in" orgone and eventually were labeled orgone accumulators. Experiments proved that the greater the layering of the walls, the greater—up to a point—the concentration or amount of orgone. In time, accumulators of up to twenty layers were built.

The next step in the discovery of the orgone came from observations of the stars and sky made through an open wooden tube.

Looking at a patch of dark sky, between stars, Dr. Reich noted the same flickering and flashing which he had observed in the specially constructed box. A magnifying glass used as an eyepiece in the tube, magnified the rays. [Gallert, p. 65]

Later, Dr. Reich called this an orgonoscope.

The flickering was less intense near the moon, and most pronounced in the darkest spots between the stars. Later Reich claimed that the same flickering was visible on white clouds in the daytime and could even be seen by looking at a white table or door, and that, being magnifiable, it was not subjective. The flickering in the sky decreases, he found, if the humidity exceeds 50 percent. This confirmed his hunch that water absorbs or draws in orgone.

Late in 1940, Reich began to notice unusual temperature effects in the orgone accumulator. Originally he had observed that the metal inside walls were "cold," but if one held his palm or tongue about four inches from the wall, one felt, after a time, a warmth and a prickling sensation. (Critics have pointed out that this sensation could arise from alterations in the circulation of the blood, when the hand is held "against gravity." To control for this, one could try holding one's hands four inches from an ordinary plaster wall.)

Then, when a thermometer was placed above the top of the accumulator, it registered a temperature higher by .2 to 1.8 degrees Centigrade than that of a thermometer in the room containing it. (Thus began attempts to prove the existence of the orgone as a physical energy. Later experiments, many carried on after Reich's death, measured the temperature inside an orgone accumulator or box—i.e., TO—compared with that inside an identical box not lined with metal—i.e., T. The formula to express this difference, TO − T, usually yielded a difference of up to one degree Centigrade.)

> Difference in atmospheric humidity appears to account for the *variation* in the temperature difference, the lowest differential occurring on humid days, and the highest differential on dry days. . . . The theory, propounded by Reich to explain this temperature differential phenomenon, is that stoppage of the kinetic energy of particles results in a temperature rise (conversion of kinetic energy into heat) . . . The metal radiates the energy both outside and inside. That radiated to the outside is absorbed by the organic outer layer, while that radiated by the metal to the inside moves freely. *Thus, movement of energy toward the inside is free, while toward the outside it is being stopped.* [Gallert, p. 66]

Reich is vague as to what actually stops the orgone going to the outside; the organic material theoretically absorbs it, but this hardly seems to equal a stopping leading to a heat effect. Also, and more significantly, later experiments seemed to find a temperature difference *inside* the accumulator. This time it was explained by the particles hitting its metal walls which, according to the previous quotation, are supposed to conduct or reflect it. In general, Reich theorized that it was stoppage of kinetic energy that caused the temperature rise, but his explanation of the actual process leaves something to be desired.

Theoretically, the

> orgone energy enters the accumulators but cannot leave them; therefore, the concentration of orgone inside the accumulators

piles up, until inflow is neutralized by leaks through ventilation spaces. Boxes built only of insulating material show no temperature difference between inside and outside atmosphere; only boxes lined on the inside with metal show this difference in temperature. [Gallert, p. 66]

At this point in his researches, Reich believed that he had sufficient evidence of a totally new and revolutionary energy to warrant sharing it with the world. But the big question was how to get people, and especially scientists, to examine it. Living now in the United States, he hit on the idea of getting the endorsement of Albert Einstein. He wrote to him and paid him a long visit on January 13, 1941, at Princeton. The story of his correspondence and rejection by Einstein is documented in a special publication of his Orgone Institute called *The Einstein Affair*. Due to both its dramatic and sociological significance I believe it deserves a detailed description. The following paragraphs summarize much of the long-drawn-out exchange between Reich and Einstein. In September, 1943, Reich wrote a description of his exchange with Einstein in a letter to his old and probably dearest friend, A. S. Neill, the founder of Summerhill in England.

Subsequent to the early correspondence and then Einstein's refusal to answer letters after February 7, 1941, Reich continued for several years to write him, on occasions, partly to keep him informed on both new experiments and new theoretical developments. Halfway through a long letter dated May 1, 1941, he says: "I know this is all extremely condensed and sounds 'crazy.' "

In that letter he stated that after discovering in 1940 the light phenomenon (orgone) in the atmosphere, he then built an accumulator. In testing this device he found that the temperature (measured by centigrade thermometer) above the accumulator was consistently higher than that of the air inside it or surrounding it. He realized the importance of this finding, since it contradicted the physical law that differences in temperature equalize; since this did not happen, the additional heat must have come from another special source. In addition, he said that the temperature difference not only was continuous but changed exactly with the weather. When the sun was shining the difference was high, up to two degrees, and when there was rain, it disappeared almost completely. "The curves . . . obtained were completely in accordance with electroscopic measurements of the energy concentration within the accumulator" (*Einstein Affair*, p. 21).

With this body of evidence, he wrote to Einstein, and a meeting was arranged for an afternoon in January, 1941. Reich told Neill that they subsequently spent nearly five hours together that day, with Reich explaining his orgone theory to Einstein. Needless to say, this kind of research was completely new to Einstein; he had not even heard of Reich's articles in the journal. Reich did not dare to advance his findings on the temperature differences until first producing an orgonoscope (an instrument Reich had devised to see the flickering of the orgone) and showing Einstein how to use it after putting the lights out. After twenty minutes or so, Einstein with genuine amazement admitted that he saw the orgone radiation, but then quickly backed down, wondering whether his eyes were playing tricks on him. Reich then told him what else he had discovered, namely the temperature differences above the accumulator. Einstein felt this to be inconceivable, but said that if it was true, it would be an explosive finding scientifically. He then offered to test this finding if Reich would send him an accumulator, and he went so far as to offer his support if it should be found to be true. Two weeks later, in the cellar of Einstein's home, the two men tested the temperature differentials, with the same results as Reich's previous ones. Einstein asked to keep the accumulator for a few weeks so as to continue his observations. In a letter to Reich about ten days later, he admitted observing the temperature changes for several days, but said that his assistant had come up with a simple explanation, i.e., that warm air was being reflected back from a cellar ceiling because of a normal convection of heat from the ceiling to the table top and so gave a higher temperature reading above the accumulator. Einstein accepted his assistant's explanation without further testing.

Reich felt compelled to retest his finding and did this in several ways, one being to place the accumulator in the open air. In these various tests the temperature difference persisted, so he felt he had disproved the assistant's explanation. However, when Reich communicated these new findings to Einstein he met only silence. This went on for many months. Apparently Einstein did not wish to commit himself any further to the inquiry.

In a long letter to Einstein dated May 1, 1941, a large section headed "The Biophysical Framework of the Temperature Phenomenon" summarized Reich's theory to that date. It is basic to future development of the whole orgone theory. He defined in this letter the term *orgone* as an energy present in all living things and even in the atmosphere and the soil. This energy, he claimed, was visible through an orgonoscope; it could also be measured electrically and by thermometer and was even visible as a blue-gray radiation on Kodachrome color film.

He believed that all living things inhaled and exhaled this energy into the atmosphere, that it was carried by the erythrocytes from the lungs to all parts of the human body and that it also probably accounted for the constant heat production of the body. He maintained too that the sun produced this energy and that it likely was related to variations in the earth's magnetism. All these views were the fruit of various experiments.

He wrote that he and an assistant had discovered that orgone affected the growth rates of organisms in different ways: protozoa in a moss infusion, when placed in an accumulator, developed (grew) at a slower pace than controls, *but* when placed *above* an accumulator their development was accelerated beyond the normal.

Using these experiments as departures, he went on to study mice with spontaneous malignant tumors. He began with 200 cancerous mice which spent half an hour per day in an accumulator. Compared with an "untreated" control group, the irradiated mice had a much longer life span (averaging 9.3 months) than the controls. He was very careful to state that he had no cure for cancer here, but wanted Einstein to be aware of the potential significance of these findings, since they paralleled, he noted, Einstein's conclusions on the link between matter and energy. He added that the relationship of orgone to electromagnetic energy was obscure, but it seemed that orgone acts "in the direction of the magnetic force and transversely to the electric force" (Reich, *The Einstein Affair*, p. E16).

Reich concluded this section of the letter by saying that all this very simple explanation of a biologically active energy would no doubt be vigorously resisted, since it shattered deeply held concepts and gave all-too-easy answers to obscure and unexplainable facts.

In spite of this and subsequent letters, Einstein made no attempt to really answer Reich's claims. A biography of Einstein points out that he was very stubborn—"if he reasoned something out . . . and made up his mind one way or the other, nobody could budge him. Take psychoanalysis . . . he studied it, reasoned it and found it inconclusive. After that, he refused to take it seriously" (Michelmore, pp. 54–55). Reich, now rejected, turned away from orthodox scientists, moved his lab to Rangeley, Maine, collected a few score of followers and pursued his investigations of the orgone without the advice or constraints of recognized leaders in biology or physics. The specific advantage of Rangeley was that it gave him the opportunity to study orgone in a high, dry atmosphere. In the next two or three years, as he experimented to uncover more characteristics of orgone, a central issue emerged—was this energy simply a form of electricity, specifically static electricity? A number of crucial experiments were carried out to elucidate this question, and these and Reich's conclusions were set forth in a long article published in October, 1944, in his *International Journal of Sex Economy and Orgone Research*, under the title "Orgonotic Pulsation." This article was ingeniously set up as a discussion between Reich and an electrophysicist. Since it is of central importance in revealing this further stage in his research and thinking it is necessary to summarize it carefully.

Near the beginning of the article, Reich sets forth certain simple and basic differences between electricity and orgone: (1) electricity is bipolar, while orgone is unitary; (2) the slow, wavelike motions of living tissues (produced by orgone) are at variance with the rapid angular motions of electricity; (3) electrical stimuli result in body sensations, but these are alien to the organism, they have a disturbing effect, and they are at variance with organic sensations. Finally, orgone's motion through the body is quite slow compared with the speed of electromagnetic radiation.

Reich also writes of orgone taking three energy forms. First, when one observes the air in an orgone accumulator at night, one sees a dim, diffuse light of a bluish-gray color. There are also small bluish dots which seem to fly by. The bluish dots "seem to come out of the [accumulator] walls at rhythmical intervals . . . as they move . . . they seem to contract and expand. When flying by sidewise, they take a trajectory similar to a parabola. This trajectory is interrupted by loop-like forms" ("Orgonotic Pulsation," p. 112). Thirdly, looking inside a small accumulator and using a green bulb for light, one sees both a blue-violet light and "yellowish-white rays which move very rapidly in all directions. It looks like miniature fireworks" (*ibid.*, p. 113). In short, various experiments, says Reich, indicate that the orgone can appear as "blue-grey vapors, blue-violet dots which float slowly and form loops at regular intervals and . . . rapid, straight yellowish rays" (*ibid.*).

Other experiments indicated that the orgone disturbs voltmeters and magnetic needles. For example, if a polystyrene rod is drawn through one's hair and moved past a voltmeter pointer slowly, at a distance of 2–5 cm., the pointer is deflected according to the way the rod is moved. From this, Reich jumps to the following interesting statement: "The so-called electromagnetic storms in the atmosphere at the time of increased sun spot activity have nothing to do with electrical or magnetic energy. They do deflect the needles of electrical measuring apparatus, that is, they disturb them in the same manner as you did when you brought about a deflection of the voltmeter with your body orgone" (*ibid.*, p. 120).

Then Reich elaborates three unusual claims: (1) that static electricity is different from electromagnetic forces;* (2) that it is really orgone; and (3) that the electroscope—which science says detects static electricity—is actually an orgonoscope, since it really measures orgone. (It follows that the electroscope, and not the voltmeter, is the appropriate instrument for determining the nature of biological energy processes.) (*Ibid.*, p. 122.) In this connection, Reich argues that the electricity of the early investigators is different from the elec-

---

* These are the energies released when a conductor, such as a wire loop, cuts through the lines of force of a magnetic field.

tromagnetic force discovered by Faraday and Volta, in which an energy is obtained by the movement of wires in magnetic fields. Here is how Reich states his case:

> Electricity—to stick to the term for the time being—was discovered, and produced, by the ancient Greeks and later by Gilbert, Cabeo, Guerike, Franklin and others, in *non-metallic* substances. Those substances which produce but do not conduct "electricity" they termed "electrica"; the metallic substances, which conduct but do not produce electricity, they termed "non-electrica." The good old electrical machine was based on the principle of friction between leather and glass; the electric energy was accumulated by way of points and "Leyden jars" . . . My contention is that the energy with which the ancient Greeks and the moderns since Gilbert were dealing was a basically different energy from that with which the physicists are dealing since Volta and Faraday; different not only with regard to the principle of its production, but *fundamentally* different. In reality, the ancient Greeks, with the principle of friction, had discovered the *orgone*. The *electric* current was not discovered until the times of Volta, Faraday, Coulomb, Ampère, etc. [*Ibid.*, p. 117]

A crucial experiment for this argument utilizes a disc of cellulose, a nonconductor, and an electroscope. The cellulose disc 6″ x 12″ is placed on the metal plate of the electroscope and when the disc is touched by a finger, the electroscope discharges gradually (*ibid.*, p. 119). This supposedly shows (a) that the body energy (orgone) travels through the cellulose, a nonconductor of electricity and (b) affects the electroscope. Thus, it is demonstrated that the electroscope responds to orgone and *not* to regular electricity, since cellulose, a nonconductor, could not conduct a static charge. In another experiment, the electroscope is given a charge "corresponding to 1,000 volts," but when connected with a voltmeter, the latter registers zero—thus apparently establishing that the energy registered is orgone and not electricity—and that the electroscope responds to orgone and not to electromagnetic forces. A further experiment with a polystyrene rod, charged with body orgone—via the hair—supposedly affects (causes variation in) an electroscope, if it is waved up and down near a fluorescent argon tube connected to the electroscope (*ibid.*, p. 121).

A number of other tests with the electroscope supposedly confirm its capacity to measure life energy, or orgone. For example, if the human hand is brought near a charged electroscope (the leaves being deflected) the leaves "begin to move; it goes down when I bring my hand close and it returns to its former deflection when I remove my hand" (ibid., p. 131). Again, an electroscope placed inside an accumulator discharges more slowly than when placed outside. (If the accumulator actually concentrates electricity, present-day theory would call for the electroscope to discharge more rapidly within than on the outside.) Reich also tested the discharge rate of an electroscope at different times of the day and found it to be more rapid in the early morning than between two and four, and again more rapid in the evening. He contends that this is at variance with the orthodox theory on ionization and its electroscope effects. Another experiment showed that it discharged more rapidly during heavy cloud formation, although the conventional ionization theory would again predict the opposite.

A further experiment produced more data at variance with electrical and ion theory. A polystyrene rod charged with energy from human hair is brought to within 5 cm. of a tubular fluorescent argon lamp. A small area of the lamp begins to glow. Then, if the same rod is brought in slowly toward the lamp from a distance of 30 cm., as it gets closer, "the lamp glows several times; this happens at shorter intervals as I get closer. If I hold the rod quiet at the same distance nothing happens. If I move it away from the lamp, it glows again, several times in succession" (ibid., p. 132). If the rod is moved lengthwise along the lamp, "there is an intermittent flickering . . . [it] does not seem to be the direct result of the even movement of the rod" (ibid.).

Elsewhere Reich explains in a very simple experiment why he concludes that orgone causes the same kind of flickering light he notices around the stars at night. He took a Faraday cage* (a room made of steel mesh from ceiling to floor and designed to ground all electromagnetic radiations) and drilled a

---

* It is significant that tests made of gifted clairvoyants in a Faraday cage have indicated that their scores (for example, on the Rhine cards) were significantly higher than usual, suggesting that some energy concentrated in it improved their performance.

few holes of ⅛ inch in one wall. Looking inside the cage, which he regards as a sort of miniature orgone accumulator, he saw a blue light that flickered strongly. He then cut a window some 5 inches square in the front wall of the cage and on the metal inside put a fluorescent glass plate across the opening. Through this, strong flickering and individual light-ninglike dots and lines became visible. He wrote further:

> As time goes on, we can distinguish vapors of a deep violet color which seem to emanate from the openings. The square where the radiation is visible is sharply defined against the black surroundings. The flickering is visible only within the square. The magnifying glass makes it possible to distinguish individual rays. On clear, dry days the phenomena are more intense and distinct than on humid or rainy days. [Reich, *Selected Writings*, p. 227]

While this experiment is interesting, the technically un-skilled observer is bound to wonder whether the holes cut in the wall of the Faraday cage may not result in a focusing of electromagnetic radiations, otherwise grounded, at these breaks in the metal wall—and so account, without any orgone, for the presence of the flickering.

All together, a variety of experiments led Reich to postulate a field force acting on another field force. The first excites the second. Thus, he hypothesized, rather than the electroscope being charged, as in the conventional theory, it or the argon lamp is being "excited." He concludes that the concept of friction (static) electricity should be changed to one of orgone excitation. He adds:

> So-called "friction electricity" has nothing to do with friction . . . "Drawing off" and "rubbing, using friction" is not the same thing. There are orgonotic phenomena which appear only if one draws off *gently* but not if one rubs strongly. Friction eliminates many reactions which are easily obtained by gentle stroking. [*Ibid.*, p. 115]

Unfortunately, evidence submitted for this remarkable state-ment is restricted to one or two experiments, whose controls are not rigorously described.

One's first criticism of this article and its experiments is that

Reich makes a lot of theory hang on a rather small number of unusual findings. Again we are not told how many times the experiments were repeated, under what laboratory conditions, nor by how many different persons. For instance, one might ask whether the subjects passing orgone to the electroscope first grounded themselves by putting their hands in water. However, if as indicated in various places static electricity is really orgone, such a question is not relevant. Thus, charging a polystyrene rod by gently passing it through the hair might produce friction (static) electricity, although when I personally tried gently passing the rod through the hair, a number of times it failed to affect my electroscope. Again, if we accept Ravitz' claims (see Chapter Five) of an electric field force around the body, the rod's charge at the electroscope could be electricity picked up from this bodily field force. The Reichians' answer to this is that electricity is a secondary manifestation of orgone. Again, where Reich notes a charge from the fingers touching the electroscope, accepting Ravitz' theory, this might be simply transferring a portion of the body's steady-state voltage to the electroscope.

A rather fascinating device that Reich made to replace the electroscope and confirm his orgone theory helps to clarify his general ideas regarding orgone energy. This new kind of "electroscope" is made with two thin silk threads, instead of the metal leaves of the electroscope. These are attached to a metal rod. The conduction from the metal rod to the metal knob is interrupted by a piece of hard rubber or plastic. When a polystyrene rod charged with hair orgone is brought to the knob, "there are several successive attractions and repulsions of the silk threads. The same happens when I take the rod away. The reaction reminds me of contracting frog's legs . . . In addition, you have reproduced the phenomenon of lumination in a *mechanical* form. The silk threads remain immobile when you do not move the rod. They move back and forth when you bring the rod close and when you remove it again" (*ibid.*, pp. 133–34).

From this and other experiments, Reich argues that static electricity and orgone are not bipolar; each simply displays two functions—attraction and repulsion. In short, body orgone is unipolar, though it can, depending upon circumstances, function to attract or to repulse. Here again Reich is repeating claims made by Galvani in the late eighteenth century. While one must pay tribute to the ingeniousness of Reich's imagination and experimentation with the electroscope, it is impossible for the nonscientist to accept specific conclusions until qualified persons have repeated and verified his work in acceptable controlled conditions.

Any evaluation of these early experiments on the orgone and its physical characteristics must deal with two issues: (a) the rigor and thoroughness of Reich as an experimental scientist; and (b) the kind and amount of subsequent positive verification. Not having seen Reich at work and having had only a brief look at his laboratory, I have been forced to rely on secondary sources with respect to the first question. These boil down, first, to conversations with persons of scientific training who saw Reich at work over a period of years, and secondly, to a careful inspection of his scientific writings on the orgone experiments.

Conversations with five persons close to Reich during the forties suggest that he was a devoted, thoroughgoing and highly responsible scientific worker. These people include an eminent psychologist, Dr. Charles R. Kelley; a reputable biologist, Dr. Bernard Grad; a well-recognized psychologist-researcher, Dr. Myron Sharaf; and two psychiatrists, Dr. Alexander Lowen and Dr. Ellsworth Baker. The last two are, of course, well known Reichian-type therapists, whose testimony may be biased. The others, while substantially persuaded of much of Reich's work, are not doctrinaire Reichians. Yet they all affirm his great scientific integrity and skill.

Dr. Kelley and Dr. Sharaf have written of this question and may therefore be quoted. Sharaf emphasizes that Reich was insistent on the necessity of noting and publicizing negative results. Thus, he says, "Reich at his best . . . was constantly coming up with negative findings, some of which led to important revisions of his concepts. To my mind this readiness of his is an important factor in explaining the extraordinary fruitful development of his work" (*Creative Process*, Vol. III, No. 1, p. 45). He also provides an instance where Reich reported in the *Orgone Energy Bulletin* (Vol. 3, 1951, pp. 72–75) a negative case wherein the accumulator, during highly disturbed weather conditions, gave a lower temperature reading than the surrounding air.

Kelley, in the same article, not only gives an illustration of how he disagreed with an important Reichian observation, in the weather-control experiments—and wrote about it in a Reichian journal in 1954—but also took issue with Reich in one of his own articles, "The Life and Death of Reich." Confirming his critical stance are such statements of his as these: "Reich was wrong about a good many things, of course, and I've never hesitated to criticize and take issue with him. . . . As good-faith repetitions of Reich's experiments are carried out, Reich will inevitably be proven wrong on many points, i.e., there will be negative findings" (*ibid.*, p. 46). But, with all this, Kelley declares Reich to have been a thoroughly competent researcher.

My personal response to books such as the *Function of the Orgasm, The Cancer Biopathy* and Reich's numerous scientific articles in his journals during the forties is generally favorable. He does cite negative instances, and uses these frequently to refine his theory, he uses controls in a consistent fashion, and he does record his data in a generally acceptable way. One gets the feeling, however, that occasionally there was a serious failure to proceed cautiously, one step at a time, that the controls could be more rigorous, that the experimental write-up could be more systematic, and that his enthusiasm could have weakened his objectivity. Diagrams, pictures and details that would clarify or improve a presentation are missing in certain instances. Sometimes, also, Reich seems to have used material as proof when it merely serves as illustration. For example, in his recorded case of a woodcutter whose ax cut his leg, and the pain was relieved after accumulator treatment. This is not a proof of orgone, since we need a control with the same injury who did not get orgone treatment. However, as an example of orgone's efficiency, and *where there is proof* from other independent experiments, this incident as an *illustration* becomes acceptable. But generally, Reich was very fair about the need for controls. For instance, at the end of *Cancer Biopathy* he called for a mass experiment with ten thousand accumulators. And in the early forties he was very hesitant about putting forth his findings as definitive. So, one comes away with the overall impression that Reich is really seeking the truth and that while his conceptualizing and theorizing talents may run away with him from time to time, his basic direction and procedures are scientifically sound and reliable.

As for the second issue, the quality and amount of verification carried out, the evidence, again, is not unequivocal. Published experiments negating the existence of the orgone as yet do not exist. As Kelley says in the above-mentioned article, "There is no counter-evidence to his experiments in any scientific publication" (ibid., p. 45). This was true in 1963 and is still true in 1972. But, the rub is that most scientific journals have no room for articles about the orgone, pro or con. Thus, we observe that so far published scientific verification of Reich's orgone energy has been carried out by Reichians or neo-Reichians and published solely in their journals. So the issue turns in part on how rigorous and scientific these Reichian experimenters and journals are. So far, since Reich's death there have been four journals published in English—Creative Process, Journal of Orgonomy, Orgonomic Functionalism, and Energy and Character, the first two in the United States and the others in Britain. All aspire to the style and format of regular scientific publications, although the multilithed Orgonomic Functionalism fell short in both format and article content. All present articles written by highly educated, literate and usually highly trained persons with an M.D. or a Ph.D. All, except for the Journal of Orgonomy, profess to take an unbiased or objective stance on Reich's ideas. They all include an occasional critical article and discussions of negative findings, observations or "puzzles." An interesting attempt to validate the heat-creating capacity of the orgone accumulator (ORAC) was reported in the November, 1971, issue of the above journal. Written by Dr. Richard Blasband, a Reichian therapist, and entitled "Thermal Orgonometry," it says:

> Regular observations of To − T [i.e., temperature inside an orgone box, minus temperature in the air nearby] have been carried out at the Bucks County, Pa., laboratory of the Oranur Research Laboratories for the past five years. Utilizing several different kinds of set-ups, we have been able to fully confirm Reich's findings. One of the simplest involved measurements of the difference in temperatures between the inside of a one cubic

foot, lidded, galvanized iron box and the outside air, six inches away. All thermometers were calibrated and were accurate to within + or − 0.1 degrees C. The box was thoroughly exposed to the outside air before being capped with its lid. The entire set-up was outdoors on an overcast day in January, 1969. The results are plotted on the following chart.

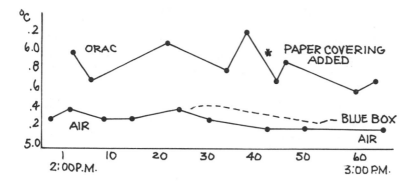

*To–T, January 19, 1969*

The following points are to be noted:

1. From 2:00 to 3:00 P.M., the period of observation, the air temperature was never more than 5.4 degrees centigrade nor less than 5.2 degrees C., while the temperature within the Orac ranged from a maximum of 6.2 degrees C. to a minimum of 5.7 degrees C: To–T was always positive, varying from a minimum of 0.3 (+ or −0.1) to a maximum of 1.0 (+ or −0.1) degrees C.

2. Any possible objection that the effects might be caused by lagging temperature changes due to insulation effects is ruled out by the following: First, the box was completely aired prior to capping it. Second, following an initial drop in both air and Orac temperatures, the Orac temperature climbed 0.8 degrees, while that of the air rose only 0.2 degrees during the same period (until 2:22 P.M.). Third, at

no time was the air temperature at first higher and then lower than that of the Orac. And fourth, no insulation was used. The excellent conductivity of heat through metal is well known.

3.  Any objection that the higher temperature within the box might be caused by electromagnetic radiation (e-m) absorption was controlled in the following ways: First, the experiment was conducted on an overcast day, thereby eliminating direct radiation from the sun. Second, we measured the temperature within a dark-blue cardboard box, which, because of its color, should have readily absorbed whatever e-m radiation may have been present. Its temperature did indeed rise higher than that of the air, but at no time was it higher than 0:15 (+ or −0.1) degrees C. This took place at a time when To–T measured 1.0 degrees C. Third, though the metal box was already fairly highly reflective, we increased its reflectivity by covering it with white paper (at the point marked* on the graph). This had little apparent effect on To–T.

We have also measured To–T indoors with similar results. A simple, yet most satisfactory accumulator for this purpose is a tin alloy coffee can wrapped in aluminum foil to increase reflectivity of light, with an outer wrapping of clear, flexible, thin plastic. The top of the can is covered with aluminum foil and capped by the plastic lid that comes with the can. The bottom is also capped. . . . One thermometer is placed through the cap of the Orac, taking care to seal it well, and the other is suspended at the exact same height above the table, about six inches away. . . .

A preliminary run without the other plastic layer yielded very satisfactory results, though the temperature variations were "spikier." In order to avoid possible spurious interpretations based on sudden large temperature changes in the room, between readings we continuously monitor air temperature with an automatic recording device. Readings following large air temperature changes are ignored. In actual practice, this is rarely necessary because of the excellent insulation of the room from the outside air. During the winter, the thermostatically controlled heaters keep the room within two degrees of the desired working temperature. In winter, despite the constant rise and fall of the room air temperature, To–T has been fairly consistently positive from 0.2 to 1.0 degrees C. Early on some mornings, presumably because of the low orgonotic excitation in the atmosphere at that time, To–T may be zero or negative. . . . Prior to precipitation, it will usually be zero or negative. . . . Prior to the onset of precipitation, even if the sky is still partially clear, To–T will usually fall, sometimes to zero or even negative values. This occurs because of the movement of energy from the earth's surface to the more highly charged clouds. With the onset of rain or snow, energy is once again released, and To–T begins to rise. [Pp. 179–83]

Altogether, unhappily, since Reich died, apart from the above article, there has been little published research by neo-Reichians on the physical aspect of the orgone.* Most articles have dealt with therapeutic or historical or philosophical issues. Quantitatively, therefore, verification has been extremely limited; qualitatively, since it has been done by friends, followers and disciples, it is open to criticism. In addition, one hears of independent individuals who have attempted a few experiments or some simple check on the accumulator and have not published their positive or negative observations. There may be dozens like this. But of published, documented and impressive articles there was still, late in 1971, a serious dearth. Early that year, Dr. Murray Melnick, a psychologist at Hofstra University, tried to get together a committee of leading American scientists, including physicists, to undertake some simple, basic replications of the orgone experiments. Everyone he approached who had scientific standing turned him down for one reason or another. In a word, thirty years after the original orgone experiments, the world lacks acceptable experimental verification or rejection of Reich's early theoretical formulations.

This does not mean that it is not possible to reach a conclusion about this theory. It means that any conclusion reached will rest on less than adequate experimental validation.

---

* A very recent article of interest entitled "The Orgone Accumulator Temperature Difference: Experimental Protocol" appeared in the *Journal of Orgonomy*, Vol. 6, No. 1 (May, 1972). The author was C. Frederick Rosenblum, a graduate student in physics at New York University.

*Chapter Two*

# ORGONE ENERGY AND THE BODY

While the discovery of atmospheric orgone provoked determined opposition, Reich's therapeutic techniques, aimed at
facilitating orgone flow in the human body, have gained considerable recognition. This is partly because they do not neces-
sitate acceptance of a revolutionary orgone theory, as the
energy so released can be, and has been, given simpler, less
ambitious labels. This, for example, is true of Alexander
Lowen's approach. He speaks of bio-energy and, while using
basic Reichian concepts in his writings, usually eschews the
term *orgone*. Other neo-Reichians have done the same thing.
Of course, this may be because Americans have understandably been anxious to avoid governmental persecution under
the judgment that had Reich's books burned and him placed
in prison. But, for Reich himself, it is orgone that becomes
blocked in its natural flow through the body—by character
armor—and acceptance of this aspect of his theory should
facilitate understanding of his over-all theory of the orgone.

It was his work in character analysis, described in his mono-
graph by that name in 1933, that led Reich directly into new
psychiatric treatment procedures, culminating in what he
called vegetotherapy. This technique, aimed at freeing up the
bio-energy (orgone) flow in the body, has been substantially
altered by Lowen and his associates in the Bio-energetic Institute.

The concept of character armoring sprang from Reich's pioneer work on patients' resistance to analysis observed in his clinical work in Germany. Reich was the first analyst to point out that almost all patients in their manner of speaking, walking and holding themselves (for example, in neck or jaw) reflect a certain character structure—defensive, self-depreciating, anxious or aggressive, and so forth. For instance, the exploiting type of character tries to manipulate others in interpersonal situations by flattery, aggressiveness or other means. Such behavior presented serious problems for therapy. Clara Thompson, in her widely respected book *Psychoanalysis: Evolution and Development* (p. 108), points out: "At first glance Sullivan's observations seem to have much in common with those of Reich about defensive character traits." Reich first attacked these resistances (defensive maneuvers) directly and found them usually reflected in specific elements of body structure, like averted or "dead" eyes, rigid chest, pulled-in pelvis, stiff jaw, and so on. He finally reasoned that the patient's specific character neurosis was directly reflected in a systematic and integrated body configuration (or armoring) that helped protect the ego from external and internal dangers. This armor meant first a reduction in the body's total psychic mobility. "He is not at ease 'within his own skin,' he is 'inhibited,' 'unable to realize himself,' 'hemmed in' as if by a wall; he 'lacks contact,' he feels 'tight enough to burst.' He strives with all his might 'towards the world,' but he is 'tied down' " (*Selected Writings*, p. 113).

He claimed that "nearly everyone by maturity has developed not only a few—or many—neurotic traits, but a way of standing, looking, holding the mouth and jaw, speaking, of breathing and holding up the chest and perhaps pulling in the pelvis, that is characteristic and set—if not rigid and largely unyielding." These visible, analyzable physical manifestations constitute the outward signs of character armor; they and the inner muscular tensions that support the eying, the breathing, the tight diaphragm or the held-in pelvis are labeled the character or muscular armoring. There are, of course, "gaps" in this armor "through which libidinal and other interests are put out and pulled back like pseudopodia . . . The degree of character armor, the ability to open up to a situation . . . constitute the difference between the healthy and the neurotic

. . . the prototype of a pathologically rigid armoring and the affect—blocked compulsive character and schizophrenic autism" (*Character Analysis*, pp. 145–46).

The neurotic character is more than a collection of neurotic symptoms. "The symptom," said Reich, "corresponds essentially to a single experience or striving; the character represents the specific way of being an individual, an expression of his total past . . . each individual character trait (e.g., exaggerated shyness) takes years to develop" (*ibid.*, p. 45). Moreover, Reich asserted a functional identity between the individual's character structure and his muscular armoring—that is, the armoring *is* the character structure in its physical form. Therefore if one can break down the armoring, one will to the same degree change the neurotic character structure. But since the rigidity of the character is locked into the body, in the armoring, it is more effective to loosen the armoring than to try to change neurotic character traits by forms of talking-out therapy like psychoanalysis.

Before describing in detail the techniques developed in vegetotherapy to loosen the character armor, the source of armor and the various forms it takes in the body should be clarified. In brief, Reich considered that the individual develops his character and expressive armoring in early childhood, perhaps beginning in the womb and then in infancy as basic drives become repressed. In modern civilization, the parental anxieties and hostilities are passed on, usually unconsciously, to the child. The insistence on excremental cleanliness and avoidance of dirt, the ban on touching the genitals and on masturbation, the demand for instinctual self-restraint (e.g., the taunt of "cry-baby") and conformity to adult norms, functional to getting children to conform to an authoritarian society, illustrate some of the basic sources of early armoring. Reich claims "the hardening of the ego"—that is, the development of armor—derives essentially from three processes.

Identification with a frustrating reality, notably with the chief person who psychically stands for this reality. Thus, if a patient characteristically held in his feelings, it is because he has said to himself, unconsciously, "I must exercise self-control as my father always told me." The second source is the aggression which the individual felt for the frustrating person but which got changed into anxiety and introjected. Such a process immobilizes much of the aggressive energy, keeping it from being expressed, and in so doing there emerges an inhibiting cast to the character.

Thirdly, according to Reich, the individual's ego develops "reactive attitudes" toward his sexual desires, using these sexual energies to contain their very expression. This means less energy is at the service of the libido, so it is less capable of asserting itself and penetrating the armor.

Thus, early prohibitions, fear of punishment and a conflicted expression of id energies help to make the armoring. So, an individual's character "serves the economic purpose of alleviating the pressure of the repressed and of strengthening the ego" (W. Reich, *Character Analysis*, p. 147). Fundamentally Reich considered that all effective blocks to natural biological movements—e.g., of curiosity, play, sex exploration, defiance of authority—where substitute actions fail to allow energy expressions, begin to build up armoring.

Raknes puts some flesh on the bare bones of this idea in the following paragraphs.

Each brake on a spontaneous, natural movement is an interference with the natural biological pulsation that we maintain to be a fundamental prerequisite and criterion of health. As long, however, as the brake on the impulse is not so bad—i.e., not so strong or not of such long duration—that the organism cannot on its own, and fairly soon, regain its free pulsation, one does not usually consider the brake of harm to its health. Fortunately, we may believe that most of the brakes to which our impulses are subjected are of this innocent kind.

"There is, however, also a series of obstacles to movement which are either so powerful or of such long duration, or of such frequency or systematic repetition, that the organism cannot on its own—at least not until long afterward—experience any impulse toward the impeded movement. In the kind of child-rearing which is still the most common, such stopping of impulses is so common that the adult organisms that have their capacity for free biological pulsation undamaged are mere exceptions" (Raknes, p. 120). The implication here is that in modern society nearly everyone is armored to a lesser or greater degree.

Depending on the kind of movement blockages experienced, and their frequency and strength, the body builds up protective armoring. Its function is to freeze energy in certain areas and assist the individual in holding in culturally wayward emotions and impulses. "Chronic muscular hypertension represents an inhibition of every kind of excitation, of pleasure, anxiety and hatred alike" (Character Analysis, p. 347). Now thoroughgoing investigation of the bodies of patients convinced Reich that the armoring was segmental—that is, that it occurred in zones or rings from the head to the pelvis. "Armor segments . . . comprise those organs and muscle groups which are in functional contact with each other, which can induce each other to participate in expressive movement" (ibid., p. 371).

The final aim of vegetotherapy is to release the armoring in the various segments, down to the pelvic area, and free up the individual so he may reach orgastic potency. Here is how he describes the segments.

While the aim of this therapy is to reactivate in the pelvis what Reich called plasmatic dreaming, involuntary sensations of pleasurable movements, one must begin the armor's dissolution at the zone farthest from the pelvis—i.e., at the head and face. Reich conceived of two segments there, one which included the forehead, eyes and cheekbone region and one that embraced the lips, chin and throat. The former ring he called ocular, the latter oral. In an armored ocular ring, if severe, there is both a contraction and stiffness—or even immobilization—of most, if not all of the muscles of the forehead, eyeball, eyelids, and even tear glands. The forehead and eyelids thus seem immobile, the eyes may protrude or seem empty of expression, so that the face could be said to look masklike. Among schizophrenics the eyes seem to be "staring into the distance." In some patients the armoring is so stiff they cannot cry; in others the eyes are narrowed to a small slit. Others seem to have a "flattened out" forehead.

Among techniques used to help dissolve ocular armoring is to get the patient to repeatedly open his eyes very wide, as if deeply frightened. Making deep questioning frowns or grimaces may also help.

Included in the oral zone are the musculature of the chin

and throat and related parts of the side and back of the upper neck. Therapy here can result in "clonisms in the lips" and a desire to cry or suck.

Clinical experience taught Reich that it is difficult to free up a lower segment—e.g., the oral—if the one above, the ocular, has not been mobilized. Indeed, effective mobilization may also be tied to freeing of *lower* as well as higher segments. Thus, actions of the oral zone, the angry biting, yelling and sucking, depend upon the freeing or mobilization of the ocular segment. What he called the gag reflex—i.e., forced attempts to vomit (sticking fingers down one's throat)—will help the patient to feel free and able to cry only if the ocular segment has been previously cleared. Crying may be very difficult also if the third and fourth segments, lower down, are seriously contracted. This is because the energy flows and emotional excitation are up and down *along* the body axis.

The third segment of armoring consists of the deep neck musculature. It gets tight from repeatedly holding back anger or crying. It may even reach up to include the tongue and its standard position in the throat. By watching the "movements of the Adam's apple" one can see how anger or crying can be literally "swallowed down" (*ibid.*, p. 375). A good way to interrupt such holding-in of emotions is to repeat the gag reflex. Its action tends to release the emotions frozen in the neck armoring.

The fourth segment includes the chest. When armored, there is shallow breathing, marked immobility of the thorax, and a persistent holding-in of breath, a stance of inspiration. Reich believed such a characterological stance was the most effective way of suppressing all kinds of emotions. He considered chest armoring central to the organism and linked to biopathic diseases like tuberculosis. Visually, the chest-armored person is seen as one who is self-controlled; he exercises restraint. The shoulders are pulled back; they· express, says Reich, a holding back. The muscles involved are the intercostal, the pectoral (large chest), the "shoulder muscles and the muscles on and between the shoulder blades" (*ibid.*, p. 376). Associated with neck armoring, an armored chest will express "suppressed spite and stiff-neckedness" (*ibid.*).

For Reich the chest segment is crucially important. He

maintains strong, natural emotions like rage, heartbreaking crying, sobbing and intolerable longing have their source in this area. And he points out that these basic emotions are alien to the really armored person: "His rage is 'cold'; crying he considers 'unmanly,' 'childish' or 'unseemly'; longing he considers 'soft,' indicating a 'lack of character' " (ibid., p. 377).

Serious chest armoring also conveys an impression that the person is "hard" and "inaccessible." One cannot get to him. Again, a certain linkage of armoring in the head, neck and chest can give the individual what is called, in a patriarchal culture, an air of distinction. He may be labeled superior, distant, "restrained" or "a stalwart character." Military training often induces such a linkage and such an impression.

Other intriguing aspects of chest armoring deserve mention. For example, Reich claims that chest armoring in women is often associated with a lack of sensitivity in the nipples. Again, people with a great sensitivity to being tickled in the intercostal region are expressing an aversion to being touched. Reich claims that chest armoring usually develops at critical junctures in the child's life, and he notes that on its dissolution in therapy there emerge disturbing memories of mistreatment, serious frustrations and disappointments in parents, et cetera.

Reich distinguishes the chest from the diaphragmatic segment (No. 5), but not in a very convincing manner. He claims that when freed up, a spontaneous pulsation occurs in this zone; but he fails to clearly spell out the character of this pulsation. He maintains that a distinctive emotional expression is bound up with this segment, but one searches in vain for a sharp description of it. For him this segment includes the diaphragm, stomach, pancreas, liver, solar plexus and "two muscle bundles alongside the lower thoracic vertebrae" (ibid., p. 380). The armoring of this ring is evident in a "lordosis of the spine" (ibid.). He notes that one can put one's hand between the patient's back and the couch. Ordinarily, too, it means that the diaphragm moves very little in regular breathing. With forced and deep breathing, it moves but slightly. Armoring in this segment means that expirations are weak; so, if the patient is asked to breathe out, he has to make quite an effort. Mobilization of this segment requires that the patient repeatedly elicit the gag reflex without at the same time interrupting his expiration. Once the diaphragm is freed, the torso tends to fold up with each expiration—i.e., the upper part tends toward the pelvis. This bending-forward, Reich claims, expresses "giving" or "surrender" in the organism.

The sixth segment is signaled by "the contraction in the middle of the abdomen" (*ibid.*, p. 388). In the back, it includes muscles running alongside the spine. When armored, the lateral muscles running from the lower ribs to the upper edge of the pelvis become hard cords, painful to pressure. Similarly, the back musculature along the spine will produce pain if palpated.

The seventh, or pelvic, segment includes practically all the muscles of the pelvis. Basically, the armored pelvis is retracted and sticks out in the back. Muscles on the outside and inside of the thigh are painful to the press. "The anal sphincter is contracted" (*ibid.*, p. 389), and the pelvis may feel dead or feelingless. This deadness will be accompanied by incapacity to feel excitations. Various pathological symptoms may accompany a "dead" pelvis: lumbago, constipation, rectal growths, inflammation of the ovaries, polyps of the uterus, irritability of the bladder or vaginal anesthesia, et cetera. The dead pelvis will hold in large amounts of anxiety or rage. If the patient is impotent, he may attempt sexual gratification by force. The pleasure sensations of the sexual act may convert themselves into those of rage, because the armored pelvis inhibits the expression of the natural involuntary convulsions and movements. A feeling of "having to get through" emerges during coitus, which may have sadistic consequences. Until the anger locked in the pelvis is released, little or no coital pleasure will be experienced.

Reich goes so far as to declare that a dead pelvis symbolically expresses contempt for the sexual act—even for the partner to it. "The rage and contempt of the sexual act is vividly expressed in the common 'cusswords' applied to it" (*ibid.*, p. 390).

Reich's systematic analysis of the armor's segments, while not completely followed by great numbers of analysts, won him some respect in psychoanalytic circles. Clara Thompson notes that Karen Horney, the author of numerous books on therapy, seems to follow Reich in describing "a complicated defensive system [which] must be uncovered" (Thompson,

p. 199). In the beginning Reich detected these defenses by observing the patient's evasive or other negative ploys in the
analytic situation. Later he was able to perceive the armoring
spots by careful inspection of the patient's posture, physical
mannerisms, way of speaking, et cetera.

A number of hints concerning the therapy Reich used—in
what he came to call vegetotherapy—are contained in the
above outline of the armored segments. First, one must begin
with the head and work downward, "clearing" each segment
in turn. Certain voluntary physical actions, like opening the
eyes wide, grimacing, or the gag reflex are helpful. In addition,
to break up the armor, Reich learned to press in certain parts
of the body (e.g., the stomach) and to squeeze certain tender
areas; this often elicits rage and anger. On this account, his
technique was facetiously referred to as "squeezing and pinching." Another practice was to encourage the patient to venti-
late hate and anger, by punching a pillow, or kicking some
likely object. In all this, the first aim is to break down and
thus mobilize the plasmatic currents felt in streamings up and
down the body. These streamings are said to be identical with
the movement of the orgone through the body (*Character
Analysis*, p. 359). The ultimate goal is the individual's full
orgasm, which indicates that the "holding back" of the armored person has changed to an easy giving of the self.

Vegetotherapy, later called orgone therapy by Reich (*ibid.*,
p. 357), consists in promoting by physiological and verbal
techniques the gradual dissolution of the armor. In many cases
setbacks occur when intense anxiety, which usually arises between therapy sessions, causes the reappearance of rigidity in
"cleared" segments. Particularly as patients feel the pleasurable plasmatic streamings, or when the pelvic segment gets
dissolved, many react with terror.

The following brief outline of orgone therapy by Dr. Mor-
ton Herskowitz helps to illustrate some aspects of its method:

> A 28-year-old female patient came for treatment complain-
> ing of abdominal pains which had been diagnosed as pyloro-
> spasm following X-ray examination. She was also troubled by
> cardiac palpitations. Her gastro-intestinal complaints were of
> three years' duration . . .

On the couch, she appeared emaciated; she held her neck stiffly; her extremities were cold; she spoke in hushed tones, and was totally unable to scream, making a helpless face with wide-open mouth and tight throat. Because of the obvious impairment of expression in the throat segment, and because her superficial neck musculature was taut and hard, the first attack was centered on her cervical area. The spastic musculature was pinched firmly and repeatedly, but she endured this painful attack silently and winced only slightly at first. As the stimuli continued, however, she gradually gave in to soft stifled cries which developed into crying, and her chest relaxed. With her now free chest movements, she experienced strong pangs of hunger.

On her next visit, she reported a two-pound gain in weight that week, after having been unable to gain a pound in years. She felt "as if a weight has been lifted." For a few days after her treatment session, she felt nausea, and this caused some anxiety, as she had always been afraid of gagging and vomiting . . .

In six subsequent weeks, she responded to irritating stimuli first by whimpering, "I'm getting angry now; I don't like to be pushed," and then by crying. As she became more readily able to cry, it became easier for her to give voice to hostile expressions such as shouting and growling. In time, she was able to include the shoulder girdle and arms in her hostile demonstrations, giving vent to hitting, scratching and screaming. This marked the end of her more obvious physical symptoms. [*Journal of Orgonomy*, Vol. I, No. 1, 2, p. 166]

A much livelier account of the early stages of orgone therapy is provided in the following excerpts from TV performer Orson Bean's article, in the same journal:

Dr. Baker sat down behind his desk and indicated the chair in front of it for me . . . he leaned back in his chair and said, "Why are you interested in working with me?" I told him that I had completed a supposedly successful psychoanalysis of ten years' duration, that I had worked hard and had gone, I felt, as far as the doctor had been able to take me, and that I felt basically unsatisfied with the results of it and with my life. I told him that I had a thriving career and a baby daughter

whom I loved, and an active sex life, although no woman
whom I cared about since the departure of my wife. I said
that I wasn't depressed or specifically unhappy, but that I
would never be satisfied until I felt fulfilled and, dare I say it,
really happy. . . .

Dr. Baker said, "I see." . . . "Well," he said, "take off
your clothes and let's have a look at you." My eyes went
glassy as I stood up and started to undress. "You can leave
on your shorts and socks," said Baker, to my relief. I laid my
clothes on the chair against the wall in a neat pile, hoping to
get a gold star. "Lie down on the bed," said the doctor. "Yes,
sure," said Willie the Robot, and did so. "Just breathe naturally," he said, pulling a chair over to the bed and sitting
down next to me. . . . "What if I get an erection, or shit on
his bed or vomit?" I asked myself. The doctor was feeling the
muscles around my jaw and neck. He found a tight cord in
my neck, pressed it hard, and kept on pressing it. It hurt like
hell, but Little Lord Jesus no crying he makes. "Did that
hurt?" asked Dr. Baker.

"Well, a little," I said, not wanting to be any trouble.

"Only a little?" he said.

"Well, it hurt a lot," I said. "It hurt like hell."

"Why didn't you cry?"

"I'm a grown-up."

He began pinching the muscles in the soft part of my
shoulders. I wanted to smash him in his sadistic face, put on
my clothes, and get the hell out of there. Instead I said, "Ow."
Then I said, "That hurts."

"It doesn't sound as if it hurt," he said.

"Well, it does," I said and managed an "Ooo, Ooo."

"Now breathe in and out deeply," he said, and he placed the
palm of one hand on my chest and pushed down hard on it
with the other. The pain was substantial. "What if the bed
breaks?" I thought. "What if my spine snaps or I suffocate?"

I breathed in and out for a while, and then Baker found my
ribs, and began probing and pressing.

I thought of Franchot Tone in the torture scene from "Lives
of a Bengal Lancer." I managed to let out a few pitiful cries
which I hoped would break Baker's heart. He began to jab at
my stomach, prodding here and there to find a tight little

knotted muscle. I no longer worried about getting an erection, possibly ever, but the possibility of shitting on his bed loomed even larger. He moved downward, mercifully passing my jockey shorts—I don't know what I had expected him to do, measure my cock or something—and began to pinch and prod the muscles of my inner thighs. At that point, I realized that the shoulders and the ribs and the stomach didn't hurt at all. The pain was amazing, especially since it was an area I hadn't thought would ever hurt. . . .

"Turn over," said Baker. I did, and he started at my neck and worked downwards with an unerring instinct for every tight, sore muscle. He pressed and kneaded and jabbed, and if I were Franchot Tone I would have sold out the entire Thirty-First Lancers. "Turn back over again," said Dr. Baker, and I did. "Alright," he said, "I want you to breathe in and out as deeply as you can and at the same time roll your eyes around without moving your head. . . . I began to roll my eyes, feeling rather foolish but grateful that he was no longer tormenting my body. On and on my eyes rolled. "Keep breathing," said Baker. I began to feel a strange pleasurable feeling in my eyes like the sweet fuzziness that happens when you smoke a good stick of pot. . . . "Alright," said Baker, "now I want you to continue breathing and do a bicycle kick on the bed with your legs." I began to raise my legs and bring them down rhythmically, striking the bed with my calves. My thighs began to ache, and I wondered when he would say that I had done it long enough, but he didn't. On and on I went until my legs were ready to drop off. Then, gradually, it didn't hurt any more and that same sweet fuzzy sensation of pleasure began to spread through my whole body, only much stronger. I now felt as if a rhythm had taken over my kicking which had nothing to do with any effort on my part. I felt transported and in the grip of something larger than me. I was breathing more deeply than I ever had before and I felt the sensation of each breath all the way down past my lungs and into my pelvis. Gradually, I felt myself lifted right out of Baker's milk-chocolate room and up into the spheres. I was beating to an astral rhythm. Finally, I knew it was time to stop. I lay there for how many minutes I don't know, and I heard his voice say, "How do you feel?"

"Wonderful," I said. "Is this always what happens?" [*Journal of Orgonomy*, Vol. 4, No. 2 (November, 1970), pp. 220–22]

The sweet fuzzy sensation of pleasure mentioned, is Bean's way of describing the plasmatic streamings. In the same article he talks of how he walked home, bursting with energy, and perceived the trees and other natural phenomena with a heightened sensibility. The next day he felt at peace with the world. "My body felt light and little ripples of pleasure rolled up and down my arms, legs and torso . . . I felt vaguely horny in a tender way" (*ibid.*, p. 223). Then late that day a sudden terror overtook him. "It was a different kind of dread . . . I felt like I . . . was starting to come apart. The anxiety was terrific and I was aware that I was involuntarily tightening up on my muscles, to hold myself together" (*ibid.*). The old armoring was reasserting itself. Describing later sessions with Dr. Baker, he tells of being forced to breathe deeply, how he screamed when pinched and later sobbed "as if my heart would break" (*ibid.*, pp. 227–28). Then he quotes the therapist: "The hard emotions have to come out first . . . the rage and the fury and the hate. Only when they're released can you get through to the tender feelings—the love and longing and sadness" (*ibid.*, p. 229).

While orgone treatment, now called psychiatric orgone therapy, can release the energy and the delightful ripples of pleasure in the body, in many cases, the therapeutic procedure is complicated and requires much training and insight to be effective. In the late forties, psychiatrists trained by Reich banded together into an association, which since his death carries on a training institute headed by Dr. Ellsworth Baker, author of *Man in the Trap*. Since Reich's death, Baker has been in charge of training doctors in medical orgonomy. In 1971, there were about twenty-five such persons affiliated with their College of Orgonomy. Baker also edits the *Journal of Orgonomy*, begun in 1967, which carries articles on case studies, researches on the orgone and reprints of some of Reich's earlier writings. In style, seriousness and professional competence, it ranks with well-known medical and psychiatric journals.

Neo-Reichians like Alexander Lowen have evolved certain

specific additions to orgone therapy. Inasmuch as they are gaining some popularity and throw some light on the orgone flow process, a brief discussion of them will be useful here. For those who desire a more extended account, reference should be made to at least two of Lowen's books, *The Physical Dynamics of Character Structure* and *Pleasure*. Throughout these he stresses the importance of breathing and of bodily movement. (See Lowen, *Pleasure*, pp. 36–43.) Lowen writes:

> Patients in bioenergetic therapy are encouraged to do special exercises which relax the muscular tensions of the body and stimulate the breathing. These exercises can also be recommended to the general public, with a word of caution to the effect that they can release feelings or produce some anxiety. They will also promote a greater self-awareness, but in this process the person may feel pain in those parts of his body that were previously immobilized. This is especially true of the lower back. Neither the released feelings nor the anxiety or pain are any cause for alarm. These exercises should not be done compulsively or pushed to an extreme, since in themselves they will not solve the complicated personality problems that trouble most people.
>
> Patients in therapy who do these special exercises designed to deepen breathing almost invariably develop tingling sensations in various parts of the body: the feet, the hands, and the face, and very occasionally over the whole body. . . . [*Ibid.*, p. 45]

It is worth noting that tingling may have a variety of sources such as hyperventilation, impaired circulation of the blood, or jarring of the nerve as in hitting the "funny bone" at the elbow. It may be related to paraesthesias or "abnormal crawly sensations," to a background feeling in the body indicating a normal experience of vitality, or be traced to the arousal of strong currents in the body caused by dissolution of an armor block.

> The basic bioenergetic breathing exercise is done by arching backward over a rolled-up blanket on a stool two feet high . . . When done at home, the stool should be placed alongside a bed so that the head and arms, which are extended backward,

hang over or touch the bed. Since this is a stress position, the mouth should be open and the breathing allowed to develop freely and easily. Most people tend to hold their breath in this position, as they do in most stress situations. This tendency must be consciously countered. [*Ibid.,* pp. 44–45]

In another basic exercise

the person bends forward to touch the ground with his finger-tips. His feet are about twelve inches apart, the toes turned slightly inward, and the knees slightly bent. There should be no weight on the hands; the whole weight of the body rests on the legs and feet. The head hangs down as loosely as possible . . .

We use this position in bioenergetic therapy to bring a person into a feeling of contact with his legs and feet. At the same time it stimulates abdominal respiration by relaxing the front wall of the body . . . When this exercise is done correctly, the legs should begin to vibrate or tremble, and they will continue as long as the breathing continues, sometimes even increasing in intensity. This vibration is normal. All alive bodies vibrate in stress positions. The vibrations of the legs and other parts of the body stimulate and release the breathing movements. [*Ibid.,* pp. 46–47]

One of the simplest and easiest expressive movements used to reduce muscular tension is kicking a bed. The person lies flat on a bed with his legs outstretched and kicks up and down in a rhythmic manner. The kicking should be done with the foot relaxed, so that the leg comes down flat on the bed. When the body is relatively free from tension, the breathing becomes synchronized with the kicking, and the head is whipped up and down with each kick as the wave of the movement passes through the body. This total bodily response will not occur if the body is held stiffly or if strong tensions prevent the head from moving. With continued exercise one learns to give in to the movement, which then becomes freer and more coordinated.

Kicking is an expressive movement. To kick is to protest, and every person has something to protest or kick about. It is a movement, therefore, which everyone can do. I recommend it to all my patients and suggest that they do 50 to 200 kicks a day. . . . They feel more alive, more energetic, and more relaxed after doing this simple exercise. . . . One may discover that his movements are awkward and uncoordinated, that he tires rapidly, and that his kicks have an impotent quality. This is an indication that a person's self-expression was blocked as a child. [*Ibid.*, p. 51]

Besides these exercises which develop the ability to express protest, anger and hate, Lowen also uses movements which help express tenderness, affection and desire.

Reaching out with the mouth and arms to kiss, to suck, to touch, and to embrace is not easy for many people. Muscular spasticities in the jaw, throat, and arms often make these movements look and feel awkward. As a result the person is hesitant about making such a movement and insecure when he does attempt it. And since a hesitant movement often evokes an ambiguous response, the person ends up with a greater sense of being inadequate and rejected. Self-confidence is the awareness that one can express himself fully and freely in any situation with appropriate and graceful movements. [*Ibid.*, p. 54]

The following excerpt from an article, "The Return of Reich," by Ann Faraday, in the British journal *New Society*, illustrates bioenergetics therapy and summarizes some of the central aims of Reich's orgone therapy:

A practical demonstration of some aspects of bioenergetic analysis was given last month by the institute's co-director, John Pierrakos, before a packed audience at the University of London School of Pharmacy. In his opening lecture, he concentrated on the intimate relation between breathing and feeling: most people's breathing is shallow and inadequate, he maintained, an inhibition which goes back to a child's habit of holding his breath to stop crying, drawing back his shoulders and tightening his chest to contain anger, and constricting his throat to prevent screaming. Deep breathing releases feelings which we are normally afraid to experience. He then showed films in which patients were enabled to release long-pent-up emotions, both of anger and fear, by controlled breathing exercises and body movements, such as kicking on the couch, beating it with clenched fists or rocking the pelvis while arched backwards over a stool. A key principle of these exercises . . . is that voluntary movements rapidly give way to involuntary ones, accompanied by the appropriate emotions.

A workshop of 20 potential British group leaders met the following weekend to receive personal initiation into these techniques by Pierrakos. He began by giving each member of the group an "instant character analysis" based on observation of facial expression, posture and body contours. Most of these were confirmed as uncannily accurate, even to the point of enabling him to give shrewd guesses at early histories underlying the person's psychological disposition . . .

Diagnosis was followed by exercises of varying kinds directed at removing the "blocks" that had been revealed, at loosening up rigidities, and at "grounding" people more firmly in responsible self-confidence by directing a flow of energy down the legs to the earth. My own experience here provided a striking example of the way in which physical action can actually bring back childhood memories. To begin with, I found (as several other people did) that deep breathing in a special position could lead to involuntary vibratory activity throughout the whole body—an intensely pleasurable sensation. After a time, as my body continued to "let go," I began to experience that "fear of orgasm" described so vividly by Reich . . .

Reich's revolutionary insight is now coming into its own, and is proving to be no mere fanatic's dream, but a vision capable of inspiring practical, balanced development towards a more joyful and healthy world.

While considerable acceptance by analysts, psychiatrists and encounter-therapy leaders for Reich's character-armoring theory exists today, some critical assessment of the theory may be helpful. First, let us note that the evidence for the immediate physical effects of orgone or bio-energetics therapy is undeniable: a lot of unknown emotion is released, the breathing improves, one feels freer and experiences ordinary pleasures more intensely. All these I have personally experienced and seen happen to many people. Hitherto locked-up energies are released. To that extent at least, a life-energy theory is confirmed. However, as with the beginning of any new system of therapy, too much has been claimed by some workers in orgone therapy. First, it has been said that (a) almost everyone has considerable armoring and (b) when it is finally broken down, a whole *new*, pleasure-focused, orgastically potent life becomes possible.

Since practically all of Reich's patients and his followers, whether it be psychiatrists like Baker or bio-energetics leaders like Lowen, are or have been middle-class, the sample is quite biased. It means we have no way of knowing precisely how widespread armoring is among lower-class, peasant or even upper-class Euro-Americans. And, of course, extending the situation panculturally, there is no clinical evidence for widespread armoring among all African or Southeast Asian peoples. Like the Oedipus complex, armoring may be only characteristic of selected classes within "advanced" patriarchal civilizations.

Again, while claims are made or implied that therapists are really healthy, orgastically (or genitally) potent persons, the evidence here is quite skimpy. Charles Kelley, who underwent extensive Reichian analysis in 1950 and knew Reich and most of the early therapists intimately, is quite convinced that

> Reich was not an unarmored genital character, nor are the doctors he trained, nor are patients concluding their Reichian therapy. There are no genital characters, not in the real world. The concept of the unarmored genital character has exactly the same kind of reality as the Scientology "clear," or Janov's "Post-Primal," who is completely devoid of defenses. [*Primal Scream and Genital Character*, p. 5]

Reichian therapists might make the following kind of comment on this:

> Premature dissolution of certain armor segments will result in chaotic situations and is disabling to the patient. This is particularly true where the eye segment, where reality is apprehended, remains blocked and the pelvis is opened up. This can happen at times in Lowen's therapy, where there is early focus on getting the legs mobilized. Other than this there is no evidence that armoring is helpful in directing and focusing energy and attention. To the contrary its presence is perceived as an

inhibiting force, a distraction, an interference in the unitary expression of the body necessary for sustained creative effort.

While Reichian therapy can be liberating and often is, its goals, as with every therapy system, are not always attained. Kelley, who tried hard to work with the leading medical orgonomists after Reich's death but was unaccepted, is highly critical of how much they have been cured of neurotic tendencies. He asks:

> And what of Reich's "angels," the "genital characters" trained by Reich (and later by a successor) to do his therapy? With the powerful techniques and the profound and germinal scientific concepts of Reich as a starting point, Reich's disciples have done not one thing to indicate that they have superior thought processes or character structures to ordinary "non-genital" characters. The Reichian movement has been characterized by unproductiveness, backbiting, jealousies and pettiness. Lawsuits and threats of lawsuits, crank letters containing attacks on each other's work and characters and motives, and other forms of factionalism have abounded. There have been only two books written by medical orgone therapists since Reich died, authored by Reich's number one and number two disciples respectively. Book number two consists entirely of an attack on book number one! [*Ibid.*, p. 6]

As a nonparticipating observer of the factional squabbles referred to, I can only testify to their presence and to the fact that they suggest that orgone therapy, while it gets at deep emotions and armoring, does not fully cure sick adults. In *Man in the Trap*, Dr. Ellsworth Baker concludes that about one in six sick adults is capable of attaining orgastic potency.

Obviously, the success of most psychotherapy depends heavily upon the character structure of the therapist and on the relationship he can establish with the patient. Where therapists are still neurotic and armored, as was clearly the case with some analysts in the early Freudian days of the twenties —let us not too deeply scan the present—patients will not be really cured. How many Reichian and neo-Reichian therapists today are truly free, unarmored genital characters one cannot say, but in my opinion there are clearly not many. These few are in tremendous demand and may be quite overworked. Their state of emotional health may easily fluctuate, for once healthy does not mean always healthy. In short, orgone therapy and therapists, while possessing powerful techniques to lessen human unhappiness and neurosis, are not owners of a panacea. These is no cure-all here. In fact, the very concept of cure may need some serious probing. A very thoughtful discussion of this whole issue is to be found in Kelley's *Primal Scream and Genital Character.**

---

* Published by Interscience Work Shop, Box 3218, Santa Monica, California 90403.

# PART
# II

*Chapter Three*

# Early History of Life-Energy Explorations

Hints of a specific life energy, functional to living organisms, come down to us from very early civilizations. Frequently they are associated with the laying of hands on the sick by priests or other communal leaders; the practice is often associated with magic. For those who seek to discount the reality of any life energy, these two linkages suffice to disqualify such practices from having any scientific interest. These critics point out that a strong suggestion of healing is conveyed by the very act of putting a hand or hands on someone's body; and that, if something therapeutic occurs, the presence of a healing energy is not demonstrated. Also, any association with magical rites and ideas, or even with religion, can also nullify the interest of many modern inquirers. Mystical or magical images are summoned up with supernatural connotations. While recognizing the reality of psychological suggestion, I believe that it is not scientifically legitimate to automatically rule out the possibility of an actual demonstrable energy. Such an energy might be present and functioning at the same time as the influence of pure suggestion or the complicated symbolic forces issuing from healing's association with magic. In a word, the widespread practice of laying on of hands and its long history *may* point to a reality beyond the purely psychological.

Scanning briefly the historical record, one finds that Chaldean priests used to lay hands on the sick. Before the Christian era, Indian Brahmans and Parsi priests laid on hands for healing. Egyptian priests well before Christ employed the same technique with success. Among the Chinese, the practice was common. In Greece, Solon was a leader in conveying some healing power by making hand strokes over the patient's body. Other ancients such as Galen, Dioscorides and Borceli made use of the idea that iron magnets, when placed near the body, had curative powers. Mesmer and others were to exploit this notion later.

The Christian Bible includes numerous incidents that point to a belief in a healing energy in the body. In the Old Testament, various stories indicate that some power is conveyed by the father's laying hands on the head of a son. The story of Elijah and the young boy (I. Kings 17:21) tells how by lying on top of the boy and praying, Elijah brought him back to life. Not only did Jesus heal—by using his hands, his eyes, or words—but his followers succeeded in duplicating many of his healings. One story, among many that suggest a healing force, is found in Luke 8:43–48; when the woman with the issue of blood touched the hem of his garment in the crowd, he said, "Who touched me, some power has gone out of me." If this story is to be believed, it would seem that his healing power had charged his garment. In the Acts of the Apostles, one reads further of healings that appear to be related to the passing on of a curative energy.

While there are many who believe that all such Biblical acts of healing were strictly of supernatural origin and prove nothing about life energies, a host of contemporary Christian and psychic healers refute this. They claim that Christ and the Apostles exercised an art open to all persons who make the effort.

Many of these contemporary healers maintain they simply tap a natural healing energy. The idea of a natural healing force* goes back at least to the time of Hippocrates—that is,

---

* For a very scholarly and detailed study of this, see Dr. Max Neuburger "The Doctrine of the Healing Power of Nature throughout the Course of Time," translated by Dr. Linn J. Boyd (unpublished Ph.D. thesis, University of Kansas Library, 1933).

to 400 B.C. "Hippocrates seems to have been the first to grasp the conception of the great healing powers of nature, and his long and wide experience made him a firm believer in those powers . . . Recovery [to him] was a work of nature" (Westlake, pp. 1–2). Could it be that a part of the natural healing force is a healing energy found in all organisms and in the air they breathe? That notion is basic to Reich's orgone theory.

A famous, but highly controversial doctor who espoused a view somewhat similar to Reich's, emerged at the time of Luther's Reformation. He was Theophrastus Bombastus von Hohenheim (1493–1541), usually called Paracelsus. It was a time of revolutionary questioning of long-standing dogmas, and Paracelsus arrogantly threw doubts on the medicine of his day. He visited many European universities, apparently secured an M.D. degree (Debus, p. 15), worked for a while as an army surgeon, traveled widely, and later became a university lecturer and personal physician to Erasmus. He devised a more effective cure for syphilis (*ibid.*, p. 17), wrote over fourteen large volumes on alchemy, ethics and philosophy, and apparently gave a new direction to medicine, by using the therapeutic capacities of chemicals. Debus points out:

> There is little agreement of modern commentators on Paracelsus. Some brand him a charlatan; others place him in the top rank of Renaissance natural philosophers. He has been described as a hero and as a villain of the Scientific Revolution. Until recently, however, few scholars have emphasized the fact that in Paracelsus and his followers there was a curious blend of the *occult* and the experimental approach to nature. [*Ibid.*, pp. 9–10]

What did Paracelsus say? The *Encyclopaedia Britannica* states that "Paracelsus founded the 'sympathetic system' of medicine, according to which the stars and other bodies, especially *magnets*, influence men by means of a subtle emanation or fluid which pervades all space." Osler says that

> Paracelsus expresses the healing process of nature by the word "munia," which he regarded as a sort of magnetic influence or force and he believed that anyone possessing it could arrest or heal disease in others. . . . In Paracelsus' own

words, "the vital force is not enclosed in man but radiates
within and around him like a luminous sphere and it may be
made to act at a distance. It may poison the essence of life
(blood) and cause disease, or it may purify it and restore the
health." [Westlake, p. 5]

Besides those who followed directly or partially in Paracelsus' medical footsteps, there were other notables who cham-
pioned his idea of healing energies. One was Pietro Pom-
ponazzi, professor of philosophy at Padua, who claimed in his
De incantationibus, which was published posthumously in
1556: "There are men endowed with the faculty of curing
certain diseases by means of an effluence or emanation which
the force of their imagination directs towards the patient. This
force affects their blood and their spirits. The healer needs
great faith, strong imagination and a firm desire to cure the
sickness." He went so far as to say that in certain circumstances this healing power could make all matter obey its
commands (Colquhoun, p. 24).

Another man to propose a theory of universal magnetism
was Jan Baptista van Helmont (1577–1644), a Belgian physician who discovered numerous chemicals. He defined magnetism as "occult influences which bodies often exert over each
other at a distance whether by attraction or impulse" (ibid.,
p. 69). He visualized a universal fluid that pervaded all nature
and was not corporeal or condensable matter and so could not
be measured or weighed. Rather, it was a pure vital spirit that
penetrates all bodies and acts upon the mass of the universe.
In humans he felt that its seat was in the blood and "that it is
called forth and directed by volition" (ibid., p. 28). He stressed
that the healer must have more power than the patient, and
that the patient's openness to the healing is very important.
In his own life Helmont used his own healing powers by laying on of hands successfully during a plague (ibid., p. 32).

Translations of some of Helmont's writings began to appear
in England as early as 1650 and profoundly affected the English Paracelsians. Many more works of Paracelsus were translated, and the Paracelsian chemists "began to take sharp issue
with the [British] College of Physicians . . . [Some] called
for the replacement of the Galenic corpus with a medicine

based on chemistry" (Debus, pp. 181–82). So, a century after Paracelsus' death, English medicine had yet to accept a new medical technique based on chemical remedies.

Robert Fludd, "the foremost English alchemical theorist of the first half of the seventeenth century" (*ibid.*, p. 90), a physician and mystic, carried on the magnetic tradition. Fludd attracted the attention of Kepler and Gassendi, and "to many Europeans seemed the most prominent of all English philosophers of his day" (*ibid.*, p. 105). He emphasized the role of the sun in health "as a source of light and life . . . the purveyor of life beams required for all living creatures on earth" (*ibid.*, p. 114). This supercelestial and invisible force, Fludd maintained, is in some way manifested in all living things, and it enters the body through the breath. Fludd also believed that the human being possesses the qualities of a magnet.

Other prominent doctors and men of science followed in the Paracelsian tradition. There was Father Athanasius Kircher, who was a German natural philosopher; Valentine Graterakes, famous as the stroking doctor, was an Irishman who later became famous in England as a magnetic healer. His counterpart, a century later, was Johann Joseph Gassner, an Austrian clergyman. Professor Glocenius wrote a book in which he attempted to explain rationally the healing miracles of Graterakes. Similarly, Burrie and Vallie in France tried to rationalize the healing successes of Gassner in Germany (Poinsot, p. 445).

Then in Italy, Luigi Galvani, one of the first discoverers of electricity, proposed an energy specific to the organic kingdom. Galvani, a physician and professor of anatomy, wrote his first major work on electricity in animals in 1791. Originally he spoke of animal electricity but later settled for the term *life force*. He claimed that animal electricity could not be the same as ordinary electricity, because it has such different characteristics. First, it is not provoked exclusively by mechanical means, but is capable "of being produced as if by nature and of manifesting itself through simple contact" (B. Bizzi, "Orgone Energy," p. 52). Secondly, it is a single "fluid not bipolar." And it is capable of two opposite functions: contraction and expansion. "Thirdly, it is not the external influence of metals which moves something static inside the living apparatus, but rather this bio-energy circulating in the living systems" (*ibid.*). In an ensuing struggle with Volta, the latter apparently proved Galvani wrong on this, through experiments which showed that the energy—e.g., that which moved a frog's legs—came from the metal.

Galvani refused to accept this verdict, however. He continued to research and to argue: "This [animal] electricity has, in common with the torpedo fish and other similar animals, the [fact that] . . . it is like a circuit from one part of the animal to the other" (*ibid.*). Galvani went further, maintaining—in Bizzi's words—that

> the circulation of this bio-energy has some kind of important and complex relationship with atmospheric electricity, whether we are dealing with the normal or the pathological. In his opinion, the skin itself provides the means of entry for the "electricity," thus playing an important role in the maintenance of an uninterrupted, dynamic balance between the organism and the "electric" ocean which surrounds it . . . On the biological level, this physical energy determines processes of actual protoplasmic expansions and contractions: expansions and contractions which we find repeated in all somatic and psychic structures: the pulsations of the cardiovascular and muscular apparatus, and of all organs in general, sympatheticotonia and vagotonia. [*Ibid.*, pp. 52–53]

Finally, in contrast to Volta, who saw the body's "organs and particularly the nerves and voluntary muscles . . . [as] simply . . . Electrometers of a new kind and of a marvelous sensitivity," Galvani's most important student and colleague, Aldini, a professor of physics at Bologna, wrote (in *Essay on the Experiments of Galvanism*):

> I flatter myself, however, that in the long run, when the necessary physiological aspects of these experiments have been pursued, they will provide better knowledge of the nature of the life forces, their different duration according to variations in sex, age, temperament, illnesses suffered and even of regions and the very constitution of the atmosphere. [Bizzi, "Orgone Energy," p. 54]

Then, coincident with the French Revolution and an era of romantic and revolutionary thinking, emerged the figure of

Anton Mesmer (1734–1815), to provide drama, innovation and excitement to the healing world with his elaborate theory of animal magnetism. Mesmer, who arrived in Paris in 1778, became a sensation almost overnight and founded a movement which endured for many decades in France and Germany. Through the intervening activities of Mesmerist Phineas B. Quimby, Mesmer was a vital force behind the theories of Mrs. Mary Baker Eddy and her Christian Science Church.* On the noncultic side, he must be credited—along with a British doctor named Braid—with the "discovery" of hypnosis, since his magnetic passes had hypnotic effects on many patients. Braid, who isolated the hypnotic side of Mesmer's work in 1842 was refused a sympathetic hearing by the medical world for twenty years (Goldsmith, p. 230). He finally got permission to read a paper at the French Academy of Science on animal magnetism and hypnosis in 1860. He stimulated Ambroise Auguste Liébault, of France, to undertake research, which led to the Nancy School of Hypnosis (Goldsmith, p. 277).

Mesmer claimed that "all things in nature possess a particular power, which manifests itself by special actions on other bodies—i.e., acting exteriorly without chemical union. All bodies, animals, plants, even stones were impregnated by this magic fluid" (Goldsmith, pp. 58–59). He had begun medicine in Vienna and found that putting a magnet over diseased parts of the body often effected a cure. Later, in working with "nervous" patients, he ran into peculiar physiological effects—for example, jerks and spasms—which led him to believe that the "magnet was mainly a conductor of a 'fluid' issuing from his own body" (Colquhoun, p. 50). He postulated a magnetic force (animal magnetism) in the human body which the magnet or "magnetic" passes activated; and he saw this force in all of nature. He claimed that animal magnetism (as distinct from mineral magnetism) can be communicated to animate and inanimate bodies in different degrees; that it penetrates all bodies; can operate at a distance; is reflected by a mirrorlike

---

* At Mrs. Eddy's first visit to Quimby, "he looked her full in the eye and then proceeded to dip his hands in water and rub her head violently in order to impart healthy electricity. Almost at once the burden of pain and weakness which she had so long borne . . . fell away and a sense of well-being rushed in and took its place" (E. M. Ramsey, p. 37). At his second treatment she said after his stroking, etc.: "She felt as if standing on an electric battery" (S. Wilbur, p. 81).

light; is invigorated by sound; and can be accumulated and transported. He also contended that it could cure nervous diseases directly and others medically, that it has properties of diagnosis and prevention, and that it indicates the degree of health possessed by individuals.

Obviously, here was a development and extension of the ideas of Paracelsus, which opponents of Mesmer were quick to point out. However, the fact that Mesmer secured some sensational cures in Paris gave his ideas the force of great originality. He quickly won the interest of Charles Deslon, a *docteur régent* of the medical faculty of the University of Paris. When, in 1781,

> the faculty resolved to extirpate the heresy by striking Deslon off the rolls, thirty of its young doctors declared themselves partisans of the new medicine. The squabble went from bad to worse, and Mesmer threatened to leave Paris. Marie Antoinette persuaded the government to negotiate with Mesmer, and to offer him "a life pension of 20,000 livres and another 10,000 livres a year to set up a clinic, if he would accept the surveillance of three government pupils." After complicated negotiations Mesmer refused, with the magniloquent gesture of a public letter to the Queen. [Darnton, pp. 50–51]

Mounting public interest in Mesmerism led to a royal investigatory commission in 1784. The verdict was almost unanimously negative with all the "cures" and effects ascribed to overheated imagination and suggestion. However, the verdict had only passing effect on the growing medicoreligious cult. By 1787 certain noted doctors were introducing animal magnetism into Germany. In France subgroups or sects emerged, especially in the provinces, where Mesmer's ideas were adopted, as they seemed to cater to local interests or circumstances.

The fascinating healing device which Mesmer innovated for medical cures was the *baquet*. This was a trough in the form of a

> circular open case about a foot high . . . at the bottom . . . on a layer of powdered glass and iron filings there lay full bottles of water, symmetrically arranged so that the necks of all

converged towards the centre; other bottles were arranged in the opposite direction with their necks towards the circumference . . . The lid was pierced with a certain number of holes whence issued jointed and moveable iron branches which were held by the patients. Absolute silence was maintained. The patients were arranged in several rows around the *baquet* connected with each other by cords passed around their bodies, and by a second chain formed by holding hands . . . Some of the patients went into convulsions, often accompanied by piercing cries, tears, hiccoughing and immoderate laughter, followed by a state of languor amounting sometimes to stupor. [Alfred Still]

A number of Reichians have written on the many parallels between the theories of "pure" animal magnetism and cosmic orgone energy. Marc Shapiro, writing in *The Creative Process* (a neo-Reichian journal), notes that

the common discovery underlying the work of Mesmer and Reich is Pulsation, which is the main action (called intensification and remission by Mesmer, expansion and contraction by Reich) of . . . the "Universal Fluid." This Universal Fluid, like Reich's orgone energy, filled all space and was the agency of all movement. It penetrated all matter, but, under laws of attraction and repulsion and the relative density of the matter, did so at different speeds and intensity. According to the resistance, fluidity, or the relative compactness of the matter through which it flowed, the fluid broke down into smaller units called *streams.* All these characteristics of Mesmer's Universal Fluid agree with those of orgone energy. [Shapiro, p. 64]

[Mesmer] felt that the nerves were the main carriers of the Fluid. . . . The nerves and body fluids carried the Fluid to all. parts of the body, where it animated and revitalized those parts. The action of the Universal Fluid, then, was the basis of life and health. To both Mesmer and Reich, health meant *wholeness.* Mesmer called it "being in harmony" with the basic laws of nature as expressed in the living organism . . . As this basic harmony is disturbed, sickness results. As the harmony is restored, health is restored. [*Ibid.,* p. 64]

The article draws other fascinating comparisons:

The most remarkable and significant relation of Mesmer's work to Reich's [was that] Mesmer discovered muscular armoring . . . Reich discovered armoring through his clinical work, and developed the discovery more completely, moving on to the discovery of the orgasm reflex, and the social implications of armoring and of orgastic impotence. But the fact is that Mesmer did discover muscular armoring in the human animal, and this played a central part in both his theoretical work and his methods of healing. Mesmer described the action of the body muscles as contraction and expansion. The ability of the muscles to undergo this action he called *irritability*. Sickness, or the conditions leading to sickness, were caused by the loss of irritability on the part of the muscles of the body. This would cause them to undergo either chronic contraction or chronic expansion. These conditions of the muscles would cause disturbances of the circulatory systems, and thence an *occlusion* of both body fluids and the Universal Fluid which animated the body. The major cause of disease and susceptibility to disease, then, was the blocking of the flow of energy through the body. The expansion and contraction, ebb and flow, intensification and remission of the Universal Fluid was the basis of living movement. This could be seen in breathing, which Mesmer called the "universal expression" of the Fluid's movement, in the action of the heart, in the excretory contractions, birth, and in "the periodical changes which we observe in sex . . . and illness." [*Ibid.*, p. 65]

Again, while Reich did not dwell on healing power in the hands, he accepted it as part of the functioning of the orgone in healthy people. Mesmer, after using magnets,

found that the best source of the Fluid for therapeutic purposes was the human body itself, particularly as conducted through the palms of the hands. For this reason, Mesmer's mode of treatment has become known generally as "laying on of hands." Mesmer felt that the flow of energy became concentrated by pointed objects, and sometimes utilized rods to concentrate the flow of energy from his body. [*Ibid.*, p. 65]

Shapiro comments on the physiological effects of sitting in Mesmer's *baquet*, and likens the effects to Reich's orgone therapy:

> The muscular spasms, the strange sensation of movement within the organism, the trembling and the convulsive movements are all typical effects observed during orgone therapy. It goes without saying that this is not all of what happens in orgone therapy—but neither, I believe, does it fully describe Mesmer's treatment. Mesmer describes people sitting around the famous therapeutic *baquet* having sudden outbreaks of laughter, anger or weeping . . . The *baquet*, then, transmitted stored energy to the patients and connected their energy flows together. [*Ibid.*, pp. 66–67]

Shapiro views the convulsive movements of persons in the *baquet* as indicating a physiological crisis in which the struggle to overcome the disease comes to a head:

> It can be conceived, in orgonomic terms, as a state of high charge, when all the body energy is involved in the final effort to shake off the disease. Reich noted the crisis in orgone therapy, when the loosening armor allows more and more energy to stream through the body, resulting in an *intensification* of the symptoms the patient originally complained of. [*Ibid.*, p. 68]

In bioenergetic therapy as practiced by Lowen, one often sees patients going through what seems to be similar highly convulsive movements.

Shapiro has produced a Table of Correspondence between the work of Mesmer and that of Reich, which sums up the gist of his argument. These correspondences are rather highly generalized and perhaps crude and inaccurate at certain points. Where this is the case, a bracketed comment has been inserted after Shapiro's point.

Table of Correspondences between the work of Mesmer and that
of Reich*

| MESMER | REICH |
|---|---|
| 1. *Universal Fluid* | 1. *Orgone Energy* |
|    a. Intensification and re-mission |    a. Expansion and contraction |
|    b. Fills all space |    b. Fills all space |
|    c. Penetrates all matter |    c. Penetrates all matter [flows through all matter] |
|    d. Carried by nerves |    d. Carried by nerves [orgone is not carried by the nerves] |
|    e. Can be stored, conducted |    e. Can be stored, conducted |
| 2. *Baquet* | 2. *Orgone Energy Accumulator* |
|    a. Becomes charged with universal fluid |    a. Becomes charged with orgone energy |
|    b. Employed therapeutically |    b. Employed therapeutically |
| 3. *Loss of muscular irritability* | 3. *Armoring* |
| 4. *Health is wholeness* | 4. *Health is wholeness* [health is indicated by orgastic potency] |
| 5. *Unhindered development of children* | 5. *Self-regulation* |

\* Marc Shapiro, "Mesmer, Reich and the Living Process."

*Chapter Four*


# MORE RECENT HISTORICAL PARALLELS
## TO THE ORGONE


Another forerunner of Wilhelm Reich was the German industrialist and scientist Karl von Reichenbach (1788–1869) who
devoted his last thirty years to investigating a universal energy
he called *od*. Reichenbach came from a bourgeois German
family, took his Ph.D. in science at Tübingen and became a
wealthy industrialist before he was forty. He was a foremost
German expert on meteorites and the discoverer of eupione,
creosote and pitticol.* He published his first findings on the
odic force, "Researches on Magnetism, Electricity, Heat and
Light in their Relations to Vital Power," in 1845 and expanded
his studies in seven other books.† For the most part, his researches were conducted in Vienna, where he found and
worked with several hundred sensitives, at least one hundred
of whom were men of scientific education.

While some of his early writings were published in scientific
journals, by 1860 Berlin's leading scientists had turned sus-
picious, considering his work but an extension of Mesmer's
controversial ideas on animal magnetism. Many of these men
were Reichenbach's friends; but, he writes, "They coupled od

---

* Eupione and pitticol are little-known chemical substances. Creosote, a
heavy, oily liquid made by distilling wood tar, is used as an antiseptic
and to reduce fever.
† These include *Der sensitiv Mensch und sein Verhalten zum Ode*.
Stuttgart, 1854.

and sensitivity . . . with the so-called animal magnetism and 'mesmerism,' and with that all sympathy was at an end" (Shepard, Introduction to Reichenbach, *Odic Force*, p. xxxv). A demonstration was arranged; but, because of the way it was handled, the sensitives were frightened and the results were unfavorable. A second demonstration was boycotted by the city's official scientists.

Reichenbach, at age seventy-nine, still persistently trying to demonstrate his 'force, had a session with the eminent Dr. Gustav Fechner.* The results were better, but not altogether conclusive, and Reichenbach died shortly afterward, without having impressed his fellow Germans. He made several "converts" among contemporary British scientists, notably a Dr. William Gregory, F.R.S.E., professor of chemistry at the University of Edinburgh and one of a long line of distinguished scholars and professors. Gregory and another British doctor separately translated and published Reichenbach's major works into English, but the books aroused no serious interest in the United Kingdom. In 1883, a small committee of the Society for Psychical Research in London, tried to rehabilitate the od, and while some of their experiments validated Reichenbach's observations, the results again were not conclusive, supposedly because of a lack of superior-type sensitives.†

Reichenbach had launched his researches when he discovered that a strong magnet, able to lift ten pounds, when passed along a person's body often produced unusual sensations, including a feeling of cold air or of pulling or drawing. The emotionally disturbed were usually the most strongly affected. Some of these disturbed people saw a luminosity around the poles of strong magnets (90-pound pull) in total darkness.

Reichenbach then confirmed that a magnet could charge water, so that sensitive subjects could distinguish the taste of this "magnetized" water. It also seemed to have a slight power to attract the limbs of sensitive subjects. Further research indicated that they felt a cool, agreeable sensation when crystalline substances like fluorspar and beryl were passed near their

---

* German scientist and philosopher, who pioneered in psychophysics. Born in 1801, he authored *Element der Psychophysik*, influential in psychology.
† By "sensitives" is meant persons with clear psychic abilities.

arms. The direction of the "pass" was apparently important, for an upward pass might produce a warm, unpleasant sensation sometimes causing violent spasms. Sensitives saw "flames" above crystals in the dark. Since crystals have no magnetic fields, Reichenbach called the energy *odyle*, or *od*.

"In addition to magnets and crystals, Reichenbach discovered eight other sources of odic energy: living organisms, the sun, moon and stars, heat, friction, artificial light, chemical reactions, electrical charges and finally the material world in general" (*ibid.*, p. 67). Iron and sulphur were perceived as especially powerful sources of od. Static electricity, in contrast to a regular electric current, had associated odic phenomena. Sensitives saw a flame from a piece of wire connected to an electrostatic machine. Again, "when a sensitive held the south pole—of a 16-inch-long bar magnet—with his right hand and the north pole with his left, a strong current was felt flowing through the arm and chest" (*ibid.*, p. 74). This feeling was almost identical to the one experienced when Reichenbach held the subject's hands in the "healing" combination—e.g., his left hand on the subject's right hand. So he reasoned the body had a polarity between the right and left sides. Plants, too, showed a complex form of polarity: in general, the points of most rapid growth and the root fibers were od positive and the tips of leaves, od negative.

The general findings of Reichenbach have been summarized as follows:

1. Odyle is a universal property of all matter in variable and unequal distribution both in space and time.
2. It interpenetrates and fills the structure of the universe. It cannot be eliminated or isolated from anything in nature.
3. It quickly penetrates and courses through everything.
4. It flows in concentrated form from special sources such as heat, friction, sound, electricity, light, the moon, solar and stellar rays, chemical action, organic vital activity of plants and animals, especially man.
5. It possesses polarity. There is both negative odyle, which gives a sensation of coolness and is pleasant; and positive odyle, which gives a sensation of warmth and discomfort.

6. It can be conducted, metals, glass, resin, silk, and water being perfect conductors.

7. It is radiated to a distance and these rays penetrate through clothes, bed-clothes, boards and walls.

8. Substances can be charged with odyle, or odyle may be transferred from one body to another. This is affected by contact and requires a certain amount of time.

9. It is luminous, either as a luminous glow or as a flame, showing blue at the negative and yellow-red at the positive. These flames can be made to flow in any direction.

10. Human beings are odyle-containers, with polarity, and are luminous over their whole surface, hence the so-called aura surrounding the physical body. In the 24 hours a periodic fluctuation, a decrease and increase of odylic power, occurs in the human body. [Westlake, p. 10]

It must be added that Reichenbach, while distinguishing his vital force from heat and electricity, was not dogmatic about its ultimate character. Thus, he wrote: "Whether magnetism, diamagnetism and od may one day be reduced to a common origin, or whether they will continue to be separated by essential differences—are questions the solution of which appears to me to be distant" (Shepard's Introduction, in Reichenbach, p. xxxi).

In the judgment of scientists, the chief weakness in Reichenbach's work was that he dealt entirely with sensitives. These were people who passed the following test.

In a somewhat shaded light, or in evening candle light, a sensitive when holding his hand before his eyes at ordinary distance, and holding the finger tips against a dark background, distant a step or two, will perceive a colorless, nonluminous current, like air, so high (a German measure) going upwards, inclining towards the south following the fingers wherever they are turned; it looks like a fine flame." [Ibid., p. lii]

This is not a heat wave, he claimed, since it will occur from cold fingers; also, the same phenomenon will be seen from magnets, crystals, metals and plants.

Obviously, working with sensitives is tricky, because these

people are often exceptionally suggestible and quick to pick up just what the researcher wants them to find—and not something objective. Some, too, are quite neurotic. Reichenbach was aware of all this, and while at the beginning many of his sensitives were people of a very nervous temperament, he claims to have used only healthy and highly educated ones later on. Reportedly he was an extremely careful and cautious researcher and insisted on repeating his experiments over and over again. (Many were repeated fifty to one hundred times.)

Several of his basic experiments are very simple. Subject A holds on to a copper or iron wire which is fed into another room twenty or thirty feet away, where the sensitive is sitting, and he is required to indicate when he or she feels something at his end of the wire. (Subject A would alternately grasp and release the wire.) In this way the sensitive indicated the presence or absence of odic force, and Reichenbach measured the speed of transmission of the force, finding it quite slow (20–30 feet a second). To show the connection of the od with sunlight, another basic experiment was to place a piece of copper-wire mesh, tin foil, or a small piece of linen, woolen, cotton or silk material in the sunlight, joined to a piece of copper or iron wire and led indoors to a dark room in which a sensitive is sitting. When the mesh is moved to the shade he can then feel—*or* see—a difference (a column of flame). Again, a glass of water left five to ten minutes in the sun was charged so that the sensitive could distinguish it from an ordinary glass of water by bringing it to his lips. To sensitives, bar magnets gave off a flame at the ends, different kinds of crystals shot out luminous rays, et cetera.

Dr. William Gregory independently verified Reichenbach's basic experiments in the 1840's. Also he writes that a Dr. Esdaile, of Calcutta, India, one of the first Western doctors to perform at that time operations without anaesthesia on hypnotized patients, concluded after extensive testing that the human body has an influence capable of being transferred and that water "charged" by od can be detected by a sensitive (Gregory, pp. iii–iv).

In 1887, a Dr. Hippolyte Baraduc published in *La Chronique médicale,* a bimonthly medical journal, some initial findings in an article, "La Force vitale." He claimed, following on von

Reichenbach, to have found a force surrounding man, penetrating him, concentrating in him, then externalizing itself
from him. He believed this energy to be linked to respiration,
coming into and leaving with the breath, and to be a cosmic
energy associated with the ether. It returns to the cosmos after
being used to maintain man. Remarkably like Reich, he believed its form to be curved, actually elliptical. Reich saw its
form as arclike.

A very industrious psychic researcher, Colonel de Rochas,
in collaboration with a Dr. Luys, of the Hôpital de la Charité,
Paris, worked with "subjects who perceived the emanation
from magnets and the human body when hypnotized . . .
The human body was found to be polarized; the right half was
a North pole and the left half a South pole" (Sudre, p. 211).
In his *Exteriorisation de la sensibilité* he examined and confirmed much of von Reichenbach's work (Gasc-Desfossés,
p. 256). "Rochas also observed that the emanation was perceived by the retina, that its color varied with different subjects . . . as did its intensity and length" (Sudre, p. 212).
But, as Sudre points out, working with hypnotized subjects
leaves one open to the power of mental suggestion. He claims
that what probably happened was that "the subject perceived
clairvoyantly the theoretical ideas of the experimenter
[Rochas] as his own empirical observations" (*ibid.*, pp.
214–15).

Another parallel to Reichenbach's od was the independent
"discovery" of so-called N rays by Professor Blondot, of the
University of Nancy, France, in 1903. Described in *Comptes
rendus de l'Académie des Sciences*, in 1904, and observed by
several other Nancy University scientists, these rays were
theoretically emitted by metals in a certain state of molecular
equilibrium. Blondot, who received the Lecompte prize of
50,000 francs for his discovery, calculated their wave length
and described their action on phosphorescent bodies. Magnets,
chemical reactions, plants, humans and lower animals emitted
them. Lead arrested most of them, as did water and moistened
paper. In general, many of his "findings" coincided with those
of Reichenbach on od. For instance, the maximum intensity of
radiation issued from those parts of the organism whose vital
activity was most intensive, and it was much stronger during

muscular or nervous activity. Some tests suggested that the rays were the result of chemical activity in the bodily tissues. Again, silk thread, wood or glass could function as N-ray conductors. And Blondot felt that the sun continuously emitted N rays.

Other French scientists expanded Blondot's findings. De Lepinay, in *Comptes rendus*, February 29, 1904, told how he found N rays in the musical sound waves generated by tuning forks and sirens. In the same decade, Charpentier and Blondot found the N energy stronger from a person who had rested through the night. Mental activity requiring real effort apparently resulted in greater emission of the rays.

Dr. Cleaves, a fellow investigator, considered the N rays to be in a band shorter than the shortest radio waves and longer than heat waves. She characterized them as (1) having high refractability, (2) having great penetration, (3) exciting to luminescence, and (4) increasing but not initiating phosphorescence. In short, she believed them capable of being reflected, refracted, polarized and brought to a focus. "They cannot," she adds, "be regarded as ions, electrons [or] effluvia"* (*Electrical Medical Digest*, p. 25). Unfortunately, when the basic experiments were repeated in non-French laboratories, the phenomena did not reoccur. A comprehensive demonstration before renowned Professor Robert Wood "showed it reasonable to attribute the effects to wishful thinking and to the . . . difficulty of estimating by eye the brightness of faint objects" (Price, *Science Since Babylon*, p. 88).

At about the same time another French doctor, Dr. E. Barety, published a rather voluminous book (662 pages) in which he described a so-called neuric energy and its various manifestations in man. According to him the Neuricidad emanates from the eyes, the finger-tips and with the breathing. He maintains that it is a question of radiating energy, capable of possessing even some physical effects. . . . Its effects on inanimate objects, he calls Neurocinesis. NRC penetrates even some materials of considerable density. It passes through certain colours but is absorbed by others. Water possesses an absorption capacity and it accumulates the respective energy. The

---

* Effluvia are defined as invisible emanations.

NRC can even impregnate various other substances which can
then become secondary irradiators. Certain metals have the
property of increasing the neuricity. [*Journal of Paraphysics*,
p. 10]

A distinguished twentieth-century researcher of life energies was Émile Boirac, Rector of the Academy at Dijon,
France. Boirac was a leading philosopher and psychologist,
who in 1910 and 1913 presided over the International Con-
gresses of Experimental Psychology held in Paris. He won the
Fanny Emden prize of 2,000 francs from the Académie des
Sciences de Paris for his submission on psychic energies, subsequently published as *Our Hidden Forces*.

The basic hypothesis which Boirac investigated was that the
human organism can exert from a distance an action analogous
to physical forces like heat, light, electricity and magnetism,
upon other human bodies, and even upon material bodies. He
hoped that insights into this force would throw much light
upon spiritoidal (spiritistic) phenomena. Working with sensitive individuals who were very susceptible to "magnetizing"
passes or thoughts and to weather changes, Boirac—who possessed magnetic power in his own hands—conducted controlled experiments of various types. Most of these are easy to
reproduce. In a simple, basic experiment, Boirac found that
with a sensitive subject the magnetizer could, by holding his
hand a foot or two away from the subject's elbow, induce in
it jerks, contractions and other movements. This was possible
when the subject's back was turned and he was engaged in
conversation with someone else. Another subject with no
verbal suggestion could be hypnotized completely just by the
right hand being placed before the forehead three times for
thirty seconds each and brought back (de-hypnotized) by a
similar "application" of the left hand. From these and other
experiments, Boirac concluded, "In my frank opinion, magnetic
radioactivity or nerve radioactivity exists as palpably as the
radioactivity of light and heat" (Boirac, p. 89).

Other experiments led Boirac to conclude that this force is
polarized and has effects which act internally within the body.
These he called curative magnetism. He quotes another French
researcher:

> Dr. Liébault [who] . . . after numerous experiments with young children came to the conclusion that a human being could, solely by its presence, exert a healthful or an unhealthful action over another, independent of any suggestion . . . "Effects very similar in nature" have been seen on animals. Certain plants having been treated by means of passes, have shown a distinct increase of growth. Other plants in a perishing state have been revived. [A. Bue, *Le Magnétisme curatif*]

Boirac also found that water gained curative properties when acted upon by passes.

A number of experiments which involved placing the hand, with straightened fingers, two to eight inches from different parts of the body of a sensitive subject, for a period of approximately thirty seconds, led Boirac to these conclusions:

> 1. The nature of these effects varies with the subjects. With the more sensitive, something always happens, although not constant in recurrence; in others, the effects produced are constant in nature. 2. Everything happens as if the human organism generates, at least with some people, an influence of unknown nature, capable of action at a distance on the organism of other individuals. 3. Most individuals appear to be good conductors of this influence . . . this influence would appear to accumulate and become stored, for a certain length of time, in parts of the body where it had been directed, with results of a more or less marked character. [Boirac, p. 163]

Some of the specific effects noted were: superficial anaesthesia; rather violent contractions in the muscles pointed at; movements of the subject's body toward the magnetizer's hand; sensations of heat, prickling, tingling or stiffness in certain body parts. Boirac claimed the effects produced by the left and right hand differed, and were discernible even when a wire was used to conduct the energy; also that the emanation was particularly strong from the fingers. A great many investigators have independently come to share Boirac's conclusions; some will be mentioned later in this chapter; others who carry out magnetic or psychic healing will be described in Chapter Nine.

As for spiritistic phenomena, Boirac maintained that

in any human "chains" formed by spiritistic sitters (holding hands or the equivalent), some are found to emanate the force, others play the part of simple conductors, while others accumulate and transform it. If among them there really be a medium, his production of energy will in a certain manner be multiplied by the quantity of force constantly received from the circle. [*Ibid.*, p. 255]

Boirac concluded that there is an influence emitted by some organisms, which he called "radiating conductors," capable of acting upon others, which he called "nonradiating insulators," and also capable of propagating itself through the organism of a third class, "nonradiating conductors." A fourth class of persons, the "radiating insulators," included subjects who could produce physical (mediumship) phenomena.

This classification of conductors helps clarify the role of psychic "fluid" in various physical effects. In hypnotism and suggestion, Boirac concluded, the fluid remains enclosed in the organism; in animal magnetism it is conveyed from one organism to another with the passes; and finally in materialization phenomena, as in the séance room, it is completely externalized.

Confirmation of some of Boirac's findings followed from the researches of S. Alrutz.

Alrutz, a Swedish professor, carried out researches to ascertain if "one nervous system can exert a direct influence upon another." His work was done with subjects in a state of light hypnosis, in which there is always a possible source of error. These subjects were seated in an armchair, their heads covered with a veil and their ears stopped with wax, so that they could neither see nor hear. Their arms rested in a horizontal position on a long box of which the upper surface consisted of a sheet of glass, five millimetres in thickness. Descending or centrifugal "passes" produced complete anaesthesia of the forearm and arm. The same result was produced through metal screens. When wool or cardboard was used, little or no result was obtained. If a strip of woolen material was laid along the sheet of glass, it seemed to diminish the subject's sensitivity and no anaesthesia was produced in the part which it shielded. That these results were not constant, Alrutz attributed to variations in the hypnotic state, and to a decrease in the specific or general sensitivity of the subject.

Ascending passes restored sensitivity with a sense of discomfort. If the subject had been pricked with a needle during the period of anaesthesia, it was at this time that he felt the prick. When a hand was held over part of the subject's body it produced tingling or smarting if he was in a light trance, while if he was in the lethargic state, sensitivity returned to the parts indicated. Alrutz also obtained changes in the power of movement by means of passes, and other experimenters obtained the same results with the subjects they had trained. [Sudre, p. 226]

Numerous French investigators carried out somewhat similar investigations, some taking off from Mesmer, some from von Reichenbach, some from Boirac and others working independently. A short list of their books, many of which are long out of print and difficult or impossible to locate in North American libraries, include: Charles Édouard Guillaume, *Vitalisme curatif par les appareils, Électro-Magnétique,* Paris, 1885; H. Baraduc, *Les Vibrations de la vitalité humaine,* Paris, 1904; Dr. Ernest Bonnayne, *La Force physique, l'agent magnétique et les instruments servant à les mesurer,* Paris, 1908; Gaston Danville, *Magnétisme et spiritisme,* Paris, 1908; Édouard Gasc-Desfossés, *Magnétisme vital,* Paris, 1907; A. Laprince, *Les Radiations humaines,* Paris, 1931; *Les Radiations cosmiques aux ondes humaines,* Paris, 1931; R. Sacerdot et M. Issantier, *Les Ondes guérissent,* Paris, 1939; Comte de Marsay, *Électricité, magnétisme, radiesthésie,* Paris, 1937; Doctor Creize, *Rayonnements humains et détection,* Paris, 1948; L. Turenne, *La Médecin en face des ondes* (Paris: Omnium Littéraire, 1931); F. Vles, *Les conditions biologiques crées par les propriétés électriques de l'atmosphère,* Paris, 1940.

Meantime, in 1937, the Russian scientist Alexander Gurvich proclaimed that all living cells produce an invisible radiation. He had experimented with freshly sprouted onions and mounted the root of one in a tube, pointing it at another onion root, which had a side exposed to the first one. Counting the number of cells in the exposed area after three hours, he found it had a fourth more than the parts unexposed. He concluded that the onion root was radiating some energy. Further research suggested that it increased the rate of yeast budding by 30 percent and effected the growth of bacteria, etcetera. Among humans, he found that muscle tissue, the cornea of the eye, blood and nerves were all senders of what he came to call mitogenetic radiation. Illnesses affected the radiation.

> Dr. and Madame Magrou, of the Pasteur Institute of Paris, found that these rays could effect the growth of plant rootlets and certain bacteria. A Dr. Borodin, of Boyce Thompson Institute of Yonkers, New York, detected this energy from 56 different kinds of living matter. The rays seem to be given off most strongly by the parts of the organism which are replaced most rapidly such as the palms of the hands and the soles of the feet . . . The tips of the fingers are very strong emitters of this energy . . . The back gives off the least . . . The sex organs in both sexes and breasts in women emit these rays quite strongly. [*Electronic Medical Digest,* Summer, 1945; T. Colson, "Living Tissue Rays"]

The work of other scientists at about the same time seemed to corroborate much of this theory. In 1936, Otto Rahn, professor of bacteriology, Cornell University, noted in his *Invisible Radiations of Organisms,* that

> the invisible radiations of living organisms are of considerable physiological significance. They play a distinct part in cell division and growth. They are evident in the healing of wounds. . . . The objection to biological radiations has been strongest in this country . . . mitogenetic radiation is not a mysterious force but the result of biochemical processes. It may be surprising that radiations have not been recognized and proven conclusively long before this. The reason may be sought in its very low intensity. The best detector is still the living organism. [Rahn, p. 90]

Reich was not unaware of these investigations and quoted Gurvich in one of his early books; he made no attempt, however, to link mitogenetic radiation directly with the orgone. Subsequent investigations of Gurvich's work have apparently failed to confirm satisfactorily his basic claims.

Another European investigator whose work provides some

background to the orgone theory is Georges Lakhovsky, a French medical doctor whose books include *The Secret of Life, The Earth and Ourselves, Contribution to the Etiology of Cancer, Cellular Oscillations* and *Radiations and Waves, the Source of Our Life.* Lakhovsky's experiments began with an intensive biological investigation of the cell and its normal activity from an electrical standpoint. Pointing out that the nucleus of the cell is composed of many tubular filaments, the chromosomes, and that hundreds of smaller filaments called chondromes are present in the cytoplasm, he maintained that

> chromosomes and chondromes are sheathed in an insulating substance and contain a liquid-like serum with the same mineral content as sea water, and consequently a conductor of electricity. Thus, these filaments constitute ultramicroscopic oscillating circuits capable of oscillating electrically over a wide scale of very short wave lengths. I have demonstrated in my work that these cellular oscillating circuits, chromosomes and chondromes, vibrate electrically under the stimulus of electromagnetic waves: cosmic, atmospheric and telluric.
>
> Many internal and external influences may upset the oscillating equilibrium of these cells. For instance, a variation or change in the field of cosmic, telluric or atmospheric waves, a demineralization of the organic matter constituting the cellular substance, traumas causing the destruction by shock of the protoplasm or the nucleus. [M. Gallert, p. 251]

To neutralize the action of disturbing rays and to give the cell the necessary stimulation for a return to its normal oscillation, Lakhovsky devised a short-wave oscillator. First he experimented on cancerous plants, using low power, from 10 to 12 watts, and a limited duration of treatment. Then he developed a superior device using weakened electrostatic waves and after much research constructed "an apparatus creating an electrostatic field covering all frequencies from 3 meters to the infra-red, so that every cell can find its natural frequency and vibrate in resonance" (*ibid.*, p. 252). This was

> an oscillator of multiple wave lengths. It included all the basic wave lengths from 10 centimeters to 400 meters, that is, all frequencies from 750,000 per second to 3 billion. But each

circuit also emits many harmonics, which, with their basic
waves, their interferences and their effluvia, can reach the
scale of the infra-red and even that of visible light . . . Since
all the cells as well as the chondromes oscillate precisely at frequencies on those gamuts, they can therefore find on the scale
of such an oscillator, the frequencies which cause them to vibrate in resonance. [*Ibid.*]

Lakhovsky began using this multiple-wave oscillator in
various Paris hospitals in the twenties and had many cures,
including patients with cancer on whom X ray and radium
had failed. He said:

These patients cured six years ago had no recurrence and are in
perfect health at this date. The treatment does not attack the
microbes directly, but *reinforces the vitality of the organism*
by accelerating cellular oscillation. It is therefore the rein-
forced organism that successfully resists the microbes and all
pathogenic causes. [*Ibid.*, pp. 252–53]

In his autobiography, *Arrow in the Blue*, Arthur Koestler
tells how, as science editor of a German newspaper, he looked
into Lakhovsky's work:

He claimed that every body-cell was an electric resonator, the
nuclear chromosomes acting as vibration-circuits, surrounded
by a faint electromagnetic field of a specific frequency. Any
disturbance in the chromosomes or in the surrounding medium
would alter this natural frequency of the cell and manifest itself as illness . . . To keep the body protected against harmful radiations, Lakhovsky invented a belt consisting of an in-
sulated, open copper spiral, to be worn round the waist or as
a necklace, armlet, ankle-band, et cetera. For a while these con-
traptions (a modern version of the charms against the evil eye)
sold like hot cakes in Paris and—thanks to an enterprising
export firm—in Budapest. Needless to say, thousands of people
swore that since they wore the Lakhovsky belt they felt "as if
reborn."

All this could have been dismissed as simple charlatanry.
However, the possibility that "there might be something in it
after all" was warranted by the fact that Lakhovsky's work
had been presented to the French Academy of Sciences, with

a very warm recommendation, by Professor Arsène d'Arsonval, an eminent scholar and past president of that Academy. Professor d'Arsonval testified i.a. that geranium plants
infected with a tumor and treated by Lakhovsky had not only
recovered completely, but had grown to five feet in height,
several times their normal size . . . There were also alleged
reports from the cancer ward of the San Spirito Hospital near
Rome according to which Lakhovsky's cure had been tried
on human patients with promising results. Though I had never
heard of a new cancer cure which did not produce some initial
results, all this sounded sufficiently intriguing to warrant a
special trip to Paris to interview Lakhovsky. I met a man as
charming as he was ambiguous, who showed me his laboratory, and repeated orally all that he had claimed before in
print. I returned convinced that half of his claims were phoney,
and none the wiser with regard to the other half. [Koestler,
pp. 309–10]

What is suggestive in Lakhovsky's notions is the use of
weak electrostatic waves at different frequencies. Treatment
consisted of having the patient seated or standing in the radius
of the apparatus for five to ten minutes every other day.
Lakhovsky admitted that his oscillator could not cure all types
of cancer in all stages of development—for example, cases
where the cancerous tumor has already destroyed important
blood vessels. Yet, as photographs in his book *The Secret of
Life* showed, it did eradicate some localized cancers that were
not far advanced. The same thing occurred when Reich's orgone accumulator was used. Finally, what is interesting is that
he took seriously the electrical basis of the human organism
and the possibility of electrostatic waves affecting the vitality
of depleted organisms.

By the 1960s evidence for the electrical basis of organisms
had become overwhelming. Thus, D. S. Halacy, Jr., a writer
on orthodox science, pointed out,

the evidence of the electrical nature of life is obvious . . . for
the flow of electrons is basic to practically all life processes.
Our nerves and muscles produce and use electricity. Our glands
seem to do the same. Circulation of the blood produces electricity and it is electricity that pulses the heart to cause that

circulation. Our senses function electrically to some extent, even our taste and smell. [Halacy, D. S., Jr., *Bionics*, p. 16]

Later, he refers to a neglected study by British researcher M. C. Potter, who showed experimentally that

the disintegration of organic compounds by micro-organisms is accompanied by liberation of electrical energy . . . The electrical effects are an expression of the activity of the micro-organism and are influenced by temperature, concentration of the nutrient medium and the number of active organisms present. [*Ibid.*, p. 157]

In the United States in 1940 Dr. George Starr Whyte published a book entitled *Cosmo-electric Culture*. In it, the results of six different types of experiments seemed to disclose the existence of one or more healing and creative energies. Whether these experiments point solely to radiation effects or to the existence of a single or multiform cosmic energy was not clear to Dr. Whyte or to unbiased readers of his work.

The first type of experiment indicated that metals like tin and iron can have a facilitating effect upon plant growth. Thus, if bright pieces of tin were dangled from fruit trees by means of wires, the fruit was improved in both size and quality. Humus fertilizer stored in iron or tin containers and used on plants seemed to produce leaves with more vivid colors than those fertilized by humus stored in wooden or other nonferrous containers.

A second set of experiments seemed to show that proper grounding and directional spacing had a favorable effect upon plant quality and growth. An elaborate layout of "cosmoelectric culture condensers" was employed. The base of tree trunks was surrounded by chicken wire, and a copper wire was led from an iron pipe about two feet below ground to the base of the tree trunks with quite beneficial growth results. Some possible explanations are: a radiation may emanate from the metal with growth results; or there may be an induction of odic or other energy from the sun through a psychic fluid circuit to the various tree trunks; or the induction may be of earth magnetic energy or of earth radiations collected perhaps originally by the iron post or the chicken wire. Numerous studies of radiesthesia in Europe leave little doubt that there is a radiation from the earth, sometimes beneficial, sometimes harmful. The unanswered question is whether the results obtained by Whyte were due to the conduction of this kind of radiation or of some other energy, perhaps odic force, since the poles and the wire mesh were also kept above the ground in the sunlight.

Another experiment by Whyte apparently showed that human beings derive a therapeutic and energizing effect from sleeping on a bed grounded by a copper wire to a copper pole inside an iron pole inserted in the earth. It is significant that the copper wire leading to the copper pole was attached to the bed pad and not the wire springs, so that apparently the energy was conducted through the mattress and bed sheet, and therefore was of the order of odic force, not ordinary electricity. It is also interesting to note that between the two-inch iron pipe and the half-inch solid copper rod placed inside it, sand was pressed down and kept continually moist. This moist sand, if we accept Reich's orgone theory, is heavily charged with orgone, and so might help to charge the copper and the iron rods.

One of the foremost contributors to recent thinking about a life energy is L. E. Eeman, a friend of the late Aldous Huxley and a nonmedical healer, who has been prominent in the British Society of Dowsers. Mr. Eeman brought together most of his experiments and findings in a provocative book called *Co-operative Healing*. Other books outline his thinking on such things as relaxation and precognition. Beginning with a remarkable self-healing performed on himself during the last years of the First World War, Mr. Eeman has become recognized in Britain as quite an authority on unorthodox healing methods.

His experiments outline quite remarkably the healing power of the laying on of hands. One of his typical and most suggestive experiments is to place two or four people in what he calls a closed circuit, in which copper wires are connected from person to person in the following fashion: With each person lying down or reclining on a couch, Eeman puts below the head and base of the spine a piece of copper-wire mesh, about twelve inches square; then ordinary copper wire is led from the copper mesh under the head of one person and grasped in the left hand of the next person; another piece of copper wire is linked from the copper mesh below the base of the spine to the next person's right hand. In this so-called "closed" circuit, when each of the people concerned is able to relax substantially, a quite pronounced relaxing and then frequently a healing effect occur after the lapse of fifteen or twenty minutes. To control against suggestion, instead of the wires being led directly from the hands to either the head or base of the spine, they are first fed through a control box, operated by the investigator. Then, without notice to the patients, the direction or linkage of the wires is reversed inside the control box (so that they go from the right hand to the head and left hand to the base of the spine area of the next person). Now instead of relaxation, an intense tension occurs. After something like twenty minutes, most of the people in the circuit get so jittery that they feel obliged to jump up and break the circuit. The tension circuit leads to some quite interesting and amazing irritation effects.

Beginning with this basic pattern, Mr. Eeman devised varied experiments, increasing the number of patients and introducing into the circuit one or more persons who have minor illnesses. When someone with a fever was placed in the circuit, after twenty to thirty minutes patients lying next to the person with the fever showed signs of a fever themselves, while the feverish person felt better. In an average fever case, after about an hour of this treatment the patient is better and the other people in the circuit have lost all fever reaction symptoms.

Mr. Eeman also tried tests in which the people in the circuit were exposed to experiments in mental telepathy. In most cases their receiving capacity for messages was reportedly remarkable. In other experiments, the copper wire linking the patients was cut in two and fed through a glass of water in which some drug was placed in solution. After the patients experienced a half hour of the relaxation circuit, they seemed to feel the same effects as if the drug in the glass had been taken internally. Apparently, whatever is being conducted along the copper wire can carry the radiations of drugs, of viruses and/or bacteria.

In another set of experiments Eeman discovered that silk thread or other types of organic materials such as linen or cotton thread can serve equally well as conductors of this energy. Joint experimentations by Mr. Eeman and Dr. Westlake, of London, have indicated certain definite characteristics of this force. It appears to flow through the body, going in the left hand, out the left foot, in the right foot, and out the right hand, and in each case it flows from the extremities, that is, the fingers and toes. Thus, if one relaxes and joins the opposite fingers and thumbs of the hands, one completes a circuit and induces a marked relaxation effect. This is increased if one also joins the opposite toes of the feet by crossing the feet over. The common habit of folding the hands and/or crossing the feet, which is especially noticeable in adults, is apparently an unconscious method of closing this energy circuit and inducing a quieting effect upon the nerves. Eeman found that in order to secure a healing effect between people, opposite hands should be held (lightly), or one should face the patient so that the healer's right hand will be placed on his left side. Furthermore, in laying hands on the body the best results are obtained, according to Eeman's tests, by placing the left hand above the right hand, if they are being placed below the neck. This position is frequently adopted quite unconsciously by many spiritual healers.

Oscar Brunler, writing from California in the late 1940's seems to have hit upon an orgonelike energy. A doctor turned physicist, Brunler published his "discovery" of a dielectric biocosmic radiation in *Rays and Radiation Phenomena* in 1950. He refers first to two German researchers, Professors Korschelt* and Ziegler, who noted in the early thirties that certain metals gave off vitalizing radiation—for example, copper— while others like lead sent out devitalizing rays. Their experiments were carried out on rats and other animals.

---

* "Prof. Korschelt lived in Japan for several years, becoming acquainted with the activities of local healers—magnetizers—and ascertained that there must be something in it. He concluded that, properly speaking, the healer serves as a kind of accumulator and transformer of the cosmic energy especially of the sun, by which he has been charged. . . . If he can, he walks in a garden, exposing his body to the sun's rays in order to replenish the energy which he later on irradiates on the place to be healed" (*Journal of Paraphysics*, p. 9).

Brunler claims that the atom's neutron radiates not an electromagnetic but a dielectric biocosmic energy. He contends that the wave length of the radiation emitted from the neutrons depends

> on their kinetic energy, i.e., their mass and velocity. This very strong radiation cannot be measured with the instruments which measure electricity or magnetic charges . . . Yet the dielectric biocosmic energy . . . can explain all so-called mysterious strange or inexplicable phenomena in many spheres . . . Light is a combination of dielectric biocosmic waves . . . and of electro-magnetic waves. [Brunler, p. 31]

This energy, he claims, gives the clue to the human halo, the radiation detected by water diviners, and the radiation of our mind. "Moreover, our mind can give shape and direction to the . . . neutrons and the biocosmic energy . . . We can transfer [these] radiations from our mind to oil, salt, and other substances The blessing of oil or salt is therefore no superstition" (Brunler, p. 34).

A fascinating parallel to much of the research already described occurs in an article by E. Laub, "Man as a Magnet," in the *Journal of the British Society of Dowsers* (March, 1958). Laub refers first to a German electrical engineer, Fritz Grünewald, who reported in his study, *Physikalisch-Mediumistischen Untersuchunger*, that he had proved experimentally that man's whole body is "the carrier of an extremely complicated magnetic field composed of a large number of magnetic centers." He referred to the existence of magnetism in the hands, along one arm and also in the head. He established by experiments that the greatest magnetic intensity was situated in the hands, especially in the tips of the fingers. And he demonstrated its existence "in the case of a certain person, P.J., whose hand possessed such a strong field that iron filings arranged themselves round it in a pattern similar to that round a magnet" (Laub, p. 162).

Laub also describes the work of a pioneer geophysicist and engineer, Erich Konrad Müller, of Zurich, who published his findings in 1932. Müller is supposed to have provided objective proof on electrical lines "of the existence of an emanation —i.e., a radiation or effluvium from the living human body—illustrating its effects with photographs" (*ibid.*, p. 163). Using a device consisting of a mobile needle placed under a glass cover, he apparently showed that from the body, especially from the underside of the fingertips, "an emanation issues and produces effects similar to those of radioactivity" (*ibid.*). This emanation is visible to a sensitive eye in a darkened room and affects photographic plates. Like odic force, it is transferable by wood, wax or copper wire, penetrates paper and glass, and is conductible by silk and other nonconducting materials. Other details on this emanation are provided by Czech Dr. Rejdak and Engineer K. Krbal, in their article "From Mesmer to the Fifth Physical Interaction and Biological Plasma":

> Eng. Müller discovered that man can not only influence this mobile system but also relay its influence. This is why he used the term "emanation of the human body" for various materials which, without the presence of man, also influence the apparatus. He discovered that energy emanating from man under certain circumstances and certain conditions of health can be conducted along a wire and through piping. The emanation is stimulated after tea, tobacco, cigars or cigarettes have been taken. The best results are after a simultaneous intake of tea and nicotine . . . A new form of energy is at work here, says Mr. Müller, and emphasises its radiational wave-like character. The energy increases when muscles are active, and emanates from fingers, but also from the breath. The energy can pervade a wooden door, wall and also a metal screen. The quantity of energy emanating from a man is irregular and changes during the experiment, radiating intermittently in fits and starts, especially if it is being transmitted through a medium other than air. [*Journal of Paraphysics*, p. 11]

Such claims recall von Reichenbach's work and that of more recent investigators. More recently a Dr. H. H. Kritzinger and a Düsseldorf doctor associate not only confirmed the existence of this human emanation, but emphasized that it is strongly dependent on weather conditions and on changes in the conductivity of the air.

All these investigators point to a nonelectrical energy, since it is presumably conducted by such nonconductors as silk or wood. The great majority of them agree on many of its specific qualities. It seems to have an affinity or relationship to magnets, it is polarized, moves through the human body in specific ways, is exuded strongly from the fingertips, is associated somehow with breathing and can be passed from one person to another. There is considerable agreement, also, that its source is the sun, that it has some relationship to metals and is absorbed by water and can affect plant life.* While not all, certainly the great majority of these qualities, according to Reich, characterize the orgone. Here is an amazing area of agreement by investigators from various countries working independently of each other.

The list of those throughout history who have hypothesized or apparently experimentally established an organic or cosmic energy could be almost endlessly extended. In recent times Henri Bergson, the French philosopher, spoke of an *élan vital*, Hans Driesch, the well-known biologist, conceived of it as entelechy. William McDougall, a leading British psychologist who taught at Harvard and wrote extensively, maintained that human instincts are regulated by a hormic energy. Dr. Charles Littlefield in the thirties described it as vital magnetism. The eminent German biologist Paul Kammerer spoke of an organic-based life force, which he labeled "formative energy." Were all these men and many others wrongheaded, misguided or blinded by subjective impressions, or were they feeling after and reporting on much the same entity?

---

* A fascinating study, still in its early stages, called *A Theory of the Dynamism of Life*, by Colonel Marcus McCausland, describes this energy as Force X. In addition to the qualities noted here, which are all in his description, he adds the following: "action independent of distance; when properly directed, can part flesh, dematerialize matter, produce telepathic effects, heal; humans can detect it; can be directed by mind; can be amplified and filtered; can be used for diagnosis of all types of disease; can be refracted, combined with other energies, resonated, stored; affected by weather, cosmic flares, is pulsing, etc." Published, *Health for the New Age*, London, 1972, pp. 8–9.

*Chapter Five*

# THE ELECTRODYNAMIC FORCE FIELD

A new way of analyzing and understanding living systems—
from plants and animals to humans—gradually evolved after
1934 out of researches by Dr. H. S. Burr, of Yale University,
and various associates. These produce a host of parallels to the
orgone theory of Reich. Although not widely accepted by
orthodox biologists, the theories and instruments developed
by Burr, Ravitz and Margenau may provide either "openings"
to an understanding of the orgone or a substitute interpretation of life energies that will gain popular acceptance. So far
not many scientists have become enthusiastic about the following network of evidence; in fact, it is possible that due to
its unorthodoxy, it might be relegated to the same scientific
limbo as the work of Mesmer, Reichenbach and Eeman.

While Dr. Burr, a leading professor of biology until his
recent retirement, originated the concept of human and other
living systems having electrodynamic force fields, a considerable extension of this research was carried on in and around
William and Mary University, in Virginia, by Dr. L. J. Ravitz
and a team of colleagues. Burr "is considered one of the rank-
ing neuro-anatomists and experimental embryologists in the
world. He serve[d] as examiner in neuroanatomy and neuro-
physiology for the American Board of Neurological Surgery,
Inc., and [was] an associate member of the American Neuro-

logical Association and the Association for Research in Nervous and Mental Disease" (Symposium II, p. 2). Ravitz, a former student of Burr's, was Director of Training and Research, Eastern State Hospital, Williamsburg, and president of the Virginia Society for Clinical Hypnosis. Much of the following research and its physical and philosophical import was first revealed publicly at a Symposium in November, 1959, held at Williamsburg, Virginia, in cooperation with the William and Mary faculty. Appearing at that time as lecturers were world-renowned physicist Dr. Henry Margenau, Eugene Higgins, Professor of Physics and Natural Philosophy at Yale, and philosopher F. S. C. Northrop. The former, a man vitally interested in life energies, discussed the application of electrodynamic field theory to living matter and to quantum mechanics.* "The technical name [given by Dr. Margenau] to these fields is *quasi-electrostatic*. They behave like electrostatic fields, except that they change in time (rhythms, etc.) . . . These are NOT 'electromagnetic' fields as such" (*ibid.*, p. 6).

In a lengthy and impressively documented article, "History, Measurement and Applicability of Periodic Changes in the Electromagnetic Field in Health and Disease," in the 1963 *Annals* of the New York Academy of Sciences, Ravitz began by noting some historical roots of his and Burr's theories.

> Noteworthy among the early propositions was that of Mead's atmospheric tidal influence on the "nervous fluid" of animal bodies in 1704, which by 1747 he had equated with electricity. Related theories and experiments were published by Freke in 1752. In 1766, Mesmer plagiarized Mead's concept of atmospheric tides, to which he appended the notion of animal tides. Other contemporary exponents of electricity in living things included Mauduit and Bertholon, the latter publishing experiments in 1783 supporting evidence for an influence of atmospheric electricity on vegetation. Despite such interest,

---

* Electrodynamics is defined in *Britannica World Language* as "the branch of physics which deals with the forces of electrical attraction and repulsion and with the energy transformations of magnetic fields and electric currents." Quantum mechanics is the theory that energy is manifested by the "emission from radiating bodies of discrete particles, or *quanta*, the values of which are expressed as the product of Planck's constant $h$, and the frequency $v$ of the given radiation."

following Galvani's controversy with Fabroni and Volta in 1792 over seemingly opposite interpretations of his frog-leg experiments, both of which ultimately proved to be correct, the study of electricity was diverted to the physicists . . . The exogenous tidal portion of earlier constructs was first reformulated most accurately by Stewart, who in 1883 postulated electric conductivity of the upper atmosphere of the earth. Next, Ekholm and Arrhenius isolated 27.32156-day tropical periods in the voltage gradients of the atmosphere. *Later, Arrhenius observed that menstruation, nativity, seizures, and other biological events were partially dependent on co-operations between specific meteorological rhythms.* [Ravitz, "History, Measurement and Applicability . . ." pp. 1144–45]

As early as 1914, Barnes and Bowman, English electrical engineers, published a book covering thirty years of investigation. They found that plant and animal life had certain electrical characteristics that changed in pathology. Most outstanding was the demonstration that cancerous areas could be definitely outlined and that a definite characteristic existed in epilepsy, neurasthenia, and all inflammatory processes.

Burr, writing in 1952 in the Yale *Journal of Biology and Medicine,* points to another forerunner of his electrodynamic theory:

Hans Driesch in modern times postulated the presence in organisms of a force not amenable to examination by the current techniques of science . . . This Entelechy . . . was to be thought of as a supernatural agency outside the present horizons of science . . . Biologists crucified Driesch for his stand, refusing to admit that there were factors in the living system . . . not amenable to rigid description and measurement. Nevertheless, the results of the operation of some sort of powerful, precise forces defining normal and experimentally modified development have rarely been more compellingly described than by this great experimental biologist. [p. 67]

The inception of Burr's work is described by Ravitz as dependent upon the construction of a suitable measuring device.

Stimulated by the pacemaking studies of Du Bois-Raymond, Mathews, Ingvar, and Lund, and by the development of field physics from the experiments of Faraday through the mathematical formulations and reformulations of Maxwell, Larmor, Lorentz, and Einstein, Burr and Northrop evolved the *electrodynamic theory of life* in 1935. Since particle physics required supplementation with field concepts in the inorganic universe, it was felt that the same constructs should likewise apply to the biologic domain, where the importance of a relational factor is even more apparent.

As measuring devices prior to 1935 tended to act as current drains, *drawing the requisite power from the organism being tested*, it was first necessary to develop new vacuum-tube instruments which would not disturb the system under observation. In accord with Maxwell's electromagnetic equations, these were designed to measure electric force fields, (i.e.) pure voltage gradients *independent of current flow* and resistance changes. [Ravitz, in *American Journal of Clinical Hypnosis*, p. 137]

For those who wish a precise technical description of the device that Burr constructed to monitor the field force, here it is. It was based on

reversible, nonpolarizing, fluid-junction electrodes, feeding into amplifiers having input impedances ranging from 10 to 1,000 megohms, with sensitivities largely conditioned by those of the output meters. Such amplifiers convert high input impedances to the low-impedance requirements of the meters. High-sensitivity galvanometers are available, as are micro- or milliammeters for less exacting determinations. For continuous measurements, both photoelectric recorders and cathode-ray oscillographs have been utilized. [Ravitz, "History, Measurement and Applicability of Periodic Changes in the Electromagnetic Field in Health and Disease," pp. 1145–46]

The electrode leads are conventionally placed on the head and chest of the subjects. In effect what Burr, with help from Cecil Lane and Leslie Nims, perfected was a microvoltmeter.*

---

* This is further described by Ravitz as reversible nonpolarizing Ag-AgC1 electrodes immersed in physio saline, employing "essentially a Wheatstone bridge circuit with high input impedance, high sensitivity, and high stability, so as to measure DC potentials as pure voltage gradients independent of current and resistance changes" (Ravitz, "Bioelectric Correlates of Emotional States," p. 499).

What is measured by these instruments is not an electric flow but the steady-state energy level of the organism. Ravitz explains the aim of such instruments by reference to auto mechanics: "A voltmeter is needed to ascertain the state function [energy level] of a car battery. Similarly it is now postulated that a voltmeter is just as necessary for determining the state function [state of functioning] of any living system" (Ravitz, "History, Measurement and Applicability . . . ," p. 1171). So what Burr and Ravitz measure is the "voltage" or power of an organism.

What is being measured is quite different from the AC current picked up in the heart by the electrocardiogram and in the brain cortex by the electroencephalogram. Nor is it simple DC. Rather, as Ravitz points out:

> Electromagnetic field phenomena [comprise] relatively steady-state voltage gradients which change relatively slowly in space-time. Even at their highest value of 150,000 ohms, skin resistances cannot affect these field measurements . . . Analogously, a reservoir of water represents a potential source of horsepower whether or not water is flowing from the reservoir. When water is drawn off to run a water wheel, current is flowing. Electromagnetic field experiments measure aspects of the equivalence of the head of water (the potential source of horsepower) and hence have virtually nothing to do with the rate of water flow, that is, with current flow. Current is a function of these fields and . . . currents are more concerned with local events than electromagnetic fields which . . . measure properties of the system as a whole. [Ravitz, "History, Measurement and Applicability . . . ," pp. 1191–92]

At the same time, Ravitz declares: "Other erroneous implications have been drawn from the terms bioelectric fields and force fields, the former perpetuating dualisms imputing a special kind of electricity in living matter, the latter evoking thoughts of static rather than dynamic fields" (*ibid.*, p. 1191). In a word, he pointedly dissociates his findings from those who, like Reich, claim there is a special kind of nonelectrical energy in living systems.

In the beginning of his researches Burr investigated the electrodynamic aspects of plants and lower forms of organic life. He examined pure and hybrid corn kernels, cotton seeds, slime mold, cucurbit fruits and trees. With the corn kernels, for example, he found the energy potential in them points to hybrid vigor and can predict growth rates, weight of the ears and height of the stalk (Ravitz, "Bioelectric Correlates . . .," p. 499). In all these and other organisms he and his followers found that specific strains and species had their characteristic steady-state (relatively unchanging) voltage gradients. He then noted that ovulation in women, and certain nonhuman organisms that ovulate, can be electrically detected, *even from the fingers and even though there are no direct nerve pathways leading from ovaries to fingers* (Symposium II, p. 2). It is interesting, too, that "Burr [was] able to control chick embryo orientations throughout the earliest stages merely by introducing ordinary horseshoe magnets at varying distances from the eggs" (Ravitz, "History . . .," p. 1148).

Out of these early investigations (1935–1955), Burr and then his fellow workers postulated that all organisms are electrodynamic systems and that they respond to electric fields from both within and without the organism. They began to talk of "electric tides" in the atmosphere—e.g., from sun and moon—and their influence on the steady state of organisms. Thus with humans and other life systems steady-state potentials generally increased significantly "every 14–17 days which occurred approximately at the time of the full and new moons, usually preceding or following the lunar day by 24–72 hours. These differences were greater in winter than summer" (Ravitz, "Bioelectric Correlates . . .," p. 500). Symptoms of schizophrenic patients studied in hospitals underwent exacerbations at such times, especially at the new moon. Doctors have long known that a significant increase in births takes place at the moon syzygies.* In addition, an annual cycle (of potential) was recorded for trees, a cycle that roughly parallels the sunspot cycle; and existing evidence suggests similar trends in human subjects. In short, living organisms are electrically affected by energy changes associated with moon and sun. Other investigations revealed sensitivities among humans to

---

* The points on the moon's orbit when it is most nearly in line with the earth and the sun.

geomagnetic influences. This fascinating subject will be developed in Chapter Fourteen.

Largely through the researches of Ravitz and his Virginia associates, such as Wilpezeski, Burr's theory was extended to cover human emotions, physical well-being and psychosomatics, psychiatry, disease, aging and hypnosis. Readings from the instruments, both of polarity and of intensity were plotted, over a period of years, for more than five hundred subjects, about 10 percent of whom were measured at least daily for spans of time exceeding twelve months (Symposium II, p. 4).

One conclusion was that here was a tool for studying human emotions, not so much *as emotions*, but in terms of their energy in both magnitude and direction—that is, readings on the microvoltmeter, with a plus or minus sign. Many subjects even learned to predict their feeling states in terms of specific quantities on the voltmeter. Ravitz stresses the value of this work on emotions, claiming that here is an important screening and predicting tool for assessing an individual's emotional stability, such as those slated for space travel. Ravitz concludes that emotions can be equated with natural energy and claims that these studies highlight the inherent impotence of bifurcated psychosomatic constructs. "Emotional" and "physical" factors appear indistinguishable—an emphasis repeatedly made by Reich in the forties, as he elaborated his orgone theories and urged the functional identity of emotion and electrical charge. Both men urge viewing the human organism as a whole entity rather than "artificially fractionating him into parts" (*ibid.*, p. 5).

Electrodynamic study of physical disease states was very revealing. Research established linkages in a whole series of health disorders, from upper respiratory infections, peptic ulcers and gastrointestinal complaints through allergies, arthritis, and nonspecific psychosomatic complaints to peripheral nerve injuries, carbuncles and cancer, and *changes in state function principally involving the H (or minus or plus sign),* and usually involving field shifts in a minus direction. In short, all these disorders *tend* to be electrically associated with shrinkages in the amount of bio-energy available. Reich attributed most of such disorders, especially degenerative diseases like cancer, to energy stasis. (This last was ascribed to emotional conflicts locked into the muscular armoring.) In effect, Ravitz supports the Reichian psychosomatic analysis. It is significant that cancer, which Reich conceived as the classic expression of the shrinking biopathy, was found associated with higher minus readings—that is, a serious lack of energy. Thus, one large study of 860 women (737 cancer-free and 123 with cancer) indicated that voltage gradients, as measured between the posterior fornix* and the ventral abdominal wall, gave the majority of cancer-free subjects positive (+) readings, and all but five of the cancerous patients negative (−) readings. Another study of 118 mice with implanted cancers revealed reproducible electrometric correlates to the presence of the malignant tumors. This is significant in view of Reich's theory on cancer (Chapter Ten).

Moderate to severe headaches and seizure states were found linked to changes in state function preceded by or associated with force field shifts of *high* intensities—that is, the E measure, the polar direction of which appears to be conditioned by individuated electrocyclic variables. Thus, sudden intense increases or decreases of energy are basic here. Field shifts to *low* intensity—of individuated polar direction—were linked to drowsiness, sleep, fatigue, coma, hypnosis, anoxia and aging. Recovery from illness or allergic symptoms, besides youthfulness, were linked with field shifts in the plus direction. These findings are compatible with the Reichian theory that predicates that the body's functioning is linked to its supply of normally flowing orgone energy.

Inasmuch as the electrodynamic theory dealt with basic energies, the discovery that the aged tended to give negative (−) readings is not surprising. Sustained minus readings were also understandably associated with fatigue and feelings of lassitude, hopelessness and being "dragged out." Not only humans, but also mice and trees, showed a general pattern of

> rising voltage gradients (upward slopes) during the first third of life, a levelling off during the middle third and a decline during the final third . . . Caucasian females of the same age . . . tend to go minus (show minus readings) at a much younger age than males . . . Beyond 50 years of age, plus

---

* A point on the top of the head.

polarity indicating a fair degree of available energy has been found only about 10% of the time in both sexes. [Ravitz, "Application of the Electrodynamic Field Theory . . .," p. 140]

Especially fascinating were Ravitz's findings on psychiatric disorders and the hypnotic state:

Prehypnotic, waking-state force field gradients show small, characteristically minute-to-minute variations (oscillations). These differ in intensity, rate of change and polarity from person to person . . . and within the same individuals at different times. Such changes are, in turn, intimately bound to the varying . . . momentums of subjects in their electrocyclic patterns . . . [Actual] hypnotic induction is often accompanied by considerable increases in force field intensity. [*Ibid.*, p. 141]

For example, up to 30 microvolts. However,

the trance state itself, following induction, is characterized by a voltage decrease relative to that during induction; and the tracing, which had previously shown varying degrees of oscillations, now smoothes to that of a straight line. As compared with waking-state patterns, force field tracings after hypnotic induction show less differentiation between individuals or within the same persons at different times. [*Ibid.*]

The first public demonstrations of electrically recorded hypnosis

were conducted at the Second Annual Scientific Assembly of the American Society of Clinical Hypnosis, in Chicago. The demonstrations were made by Lt. Warren J. Elliott, Fitzsimons Army Hospital, Denver, an associate of Dr. Ravitz . . . It is estimated that the Chicago demonstrations were witnessed by several hundred psychologists, dentists, and physicians who attended the meetings. Cathode-ray tubes and photoelectric, inkwriting recorders were attached to the Burr-Lane-Nims millivoltmeter to clarify the electric picture. [Symposium II, p. 4]

Ravitz reports on investigations of the electrodynamic aspects of schizophrenia carried out at Duke University School

of Medicine in 1950. He posits a fascinating relationship between schizophrenia and energy overcharging:

> Consider that if a car battery is overcharged, it may not
> work properly and will tend to burn out. This depends on
> degree of overcharging, duration, and the inherent durability
> of the battery. When the battery is almost burned out, an ex-
> cessive charge is often needed to start the motor. Shaking it
> also helps temporarily, but the battery will not hold its charge
> for long. This analogy provides a concise picture of electro-
> magnetic field changes in schizophrenia . . . The greatest
> field strengths, *states of excitation*, are found in disturbed
> persons despite their "front" and regardless of apparent dura-
> tion of their disorders. Improvement is preceded by or corre-
> lated with sustained voltage drops. [Ravitz, "History, Measure-
> ment and Applicability . . . ," pp. 1157–58]

He emphasizes that in order to know what is really going on
with some schizophrenics—e.g., those mute or apparently exhausted—one needs to take a field profile over a given period
of time with the microvoltmeter. In short, electrodynamic
gradients assist in understanding the process of therapy.
While this is not the place to advance details, it is interesting
that this analysis of schizophrenia in terms of energy levels
accords significantly with that of Reich's orgone-theory perspective.

How scientific are the various claims of the Burr-Ravitz
group? Admittedly, until more investigators from other institutions confirm the basic findings, no final statement is possible. The present variety of researches, in different places and
by different men, is some testimony to reliability but, from
a modern scientific standpoint, not enough. The fact that a
number of Russian scientists evolved parallel types of investigations in the sixties lends weight to their claims. (The Russian researches are reported in Chapter Fourteen.) The fact
that, with individual schizophrenics, future emotional states
were accurately predicted on occasion by force field readings,
and that hospital admission flows were also predicted (*ibid.*, p.
1186) strengthens the theory's acceptability. If more predictions can be made and verified, interest will speedily develop.

Ravitz claims other strengths for the Burr theory. He summarizes four conclusions regarding the microvoltmeter meas-urements drawn from his research and that of Burr and col-leagues, as follows:

(1) a high degree of stability exists for both intensity and polarity measurements over relatively short spans of time; (2) stability of such measurements is higher when the polarity (directional) quantity is considered; (3) stability tends to de-crease somewhat as the interval between measurements is lengthened; and (4) controls (subjects) show considerably less stability than hospitalized patients with chronic conditions, and perhaps less than the mixed group comprising hospitalized patients with both semi-acute and chronic conditions. The last finding may result from the fact that controls do not tend to maintain high intensities of either phase. [*Ibid.*, p. 1151]

The last conclusion implies a considerable state of flux in the energy gradients of "normal" people. Each individual exhibits his or her own profile of diurnal, 7-day, 14-to-17-day, and seasonal fluctuations and rhythms; in sum, the living or-ganism *pulsates*, to "individually timed rhythmic variations, whose intensities, elasticities and directions are amplified, con-densed, accelerated, decelerated and reversed in accord with other frequencies" (Ravitz, "Application of the Electrody-namic Field Theory . . . ," p. 139).

Ravitz is claiming that living systems pulsate, and each at its own rate and rhythm and in interdependence with other energy systems. Compare the over-all Burr-Ravitz perspective with the following paragraphs by psychiatrist John C. Pier-rakos, the associate of Alexander Lowen of the Institute of Bioenergetic Analysis:

One of the most interesting aspects of rhythmic activity in an organism is the pulsation of its energy field. The human body is surrounded by a force field which has been described as an aura or atmosphere . . . I have observed and studied this phenomenon for about 15 years . . .

The significance of the field for our present discussion is that it pulsates. It appears and disappears at a rate of about 18 to 25 times per minute. This rate is independent of any other known bodily rhythm such as respiration, heart beat,

etc. It varies, however, with the overall degree of bodily excitation. When a patient, for example, strikes the couch repeatedly with a feeling of anger, the rate of pulsation may
increase to 40 per minute . . . When the body goes into vibration as a result of deeper breathing and strong feeling the rate
increases markedly, going as high as 45 to 50; at the same
time, the width of the field extends further and the color becomes brighter. The field reflects the level of excitation and
the intensity of feeling in the body . . .

The rate of pulsation changes with the diurnal activity of
the individual. During sleep it slows down, the width diminishes, and the color is weaker. The rate generally reaches its
peak in the afternoon . . . It is less in the early morning soon
after awakening, and falls off again in the evening when the
organism is exhausted.

The pulsatory rate of the human field is also related to a
similar field phenomenon in the earth's atmosphere. One can
observe on the horizon, over the ocean, or over the mountains
a layer of light blue that appears and disappears with a regular
rhythm . . . It can be seen by most people who will take the
trouble to look at the horizon during the early morning or late
afternoon hours.

The rate of pulsation of this field about the earth's surface
varies from 12 to about 40 per minute. It changes with the
time of day, lowest about midnight, highest about noon. On
oppressively humid days this field phenomenon is weaker and
the rate is decreased, dropping to half. The approach of a
storm also decreases the rate, but it picks up again when the
sky is clear. Seasonal variations also occur; the rate is highest
in the spring and lowest in the winter. I have noticed a correlation between the pulsation of the earth's field and that of the
human organism. On days when the earth's field is intense and
pulsating rapidly, the human field is more vivid and more active. On other days when the atmosphere is oppressive both
fields decline, the human field becomes less vivid and its rate
is slower.

People are known to be sensitive to the weather. They become excited when the day is bright and clear; they feel
apathetic when the day is dull. The excitatory processes that
occur in our bodies appear to be related, in a mysterious way,

to the excitatory processes in nature. We are governed more than we realize by the natural forces about us. The living organism is part of the larger whole in which rhythm and vibration are the unifying factors. They are also the factors that influence the relationship of one person to another. The rhythmic patterns of a person affect other people in his proximity. Living organisms are resonating systems that respond to each other's vibrations. Some people believe that they can pick up the waves that others emit, others refer to this sense of rapport and communication as being "on the same wave length." Whether such waves exist or not, the fact is that people are sensitive to other people's states of excitation in a physical way. We respond to their joys and sorrows. We are attuned to their vibrations. [Lowen and Pierrakos, *The Rhythm of Life*, pp. 32–33]

When a person speaks at his ground tone or voice at a range and register that is resonant, the energy field lights up and the pulsatory rate increases to about 35/min depending on the individual. When a speaker with vibrant voice and stimulating ideas addresses an audience, one can observe that the energy fields of the listening individuals become brighter, speed up one by one, and fuse over their heads and the group of people pulsates as a whole at a higher rate.

When a child is reprimanded by his mother in a stern voice, his field slows down and loses its brilliance. When a person shouts with anger and pounds his fists, the field becomes red, looks like it is transversed by streaks, like a porcupine's quills and doubles its pulsatory rate. In deep sobbing with convulsive discharge, the field becomes deep purple, especially over the chest, and increases its rate of pulsation.

In some experiments with plants and crystals we conducted in my office with Dr. Wesley Thomas, we found that a chrysanthemum's field contracts markedly when a person shouts at it from a distance of five feet, and loses its blue-azure color, while its pulsation diminishes to one-third. In repeated trials, keeping live plants more than two hours daily near the heads of screaming patients (a distance of three feet away) the lower leaves started falling down and the plant withered within three days and died. It appears to me that screaming is not only a mechanical vocal sound expression, but a live process affecting the whole body, stimulating the field to expand or contract and at times causing injury to the experimental plants . . .

It is obvious from these observations that the voice expresses or blocks vibratory movements from the core of the organism. What moves a person is his inner attitude. The pulsatory movements of life, of which the energy field is an expression, combine with the physical movements of the body in unitary functioning . . . One can say that the ability to express the full range of one's feelings vocally as well as verbally is a measure of health. [Lowen and Pierrakos, *Self-Expression, New Developments in Bio-energetic Therapy*, p. 12]

In these fascinating statements Pierrakos and Lowen seem to be building on Reich's insight that all living organisms pulsate and that basic to such pulsatory movement is a specific energy, the orgone.

Ravitz believes his approach may be an integrator of knowledge—that is, may lead to some unified theory of life. The same kind of vision haunted Reich. The key speakers at Ravitz' 1959 symposium at Williamsburg—Northrop, Burr and Margenau—dealt with "Field Theory as an Integrator of Knowledge." Facts from biology, psychology, mathematics, medicine and physics, they hinted, seem to become coordinated through the electrodynamic-field theory. Margenau includes quantum mechanics as a topic that may also fit in. In spite of the complexity of the human biological system and its sensitivity to the total field of electrical forces, which in their turn are influenced by the individual biological units and their pulsations, Ravitz sees the electrodynamic perspective as ultimately unifying and highly meaningful. The selfsame vision, for much the same reasons, inspired Reich to develop, apply and extend his orgone theory. It is possible that to some extent Ravitz and Reich were covering much the same ground and may have been speaking of the same energy, but in different conceptual terms. Certainly both came to realize the astounding interdisciplinary potentialities of their findings.

# PART
# III

*Chapter Six*

# INDIAN PHILOSOPHIES
## AND THE ORGONE

While scientific analysis and values still dominate the thinking of the middle classes in the West, increasing numbers of young people are looking to the East for philosophic guidelines. The ethnocentrism that paints the West as superior in all important respects to Oriental culture and ideas is beginning to weaken. It is valuable, therefore, to show how Hindu, theosophical and occult traditions, which all speak of life energies—enclosed in and transcending the human body—throw light on the quest for the orgone. What follows has nothing to do with scientific experiments, but it should intrigue the open-minded and underline the urgency of understanding the energy base of life.

Prana, in Hindu and yoga thinking, is a universal energy more basic than atomic energy. A recognized text says:

> Heat, light, electricity, are all manifestations of prana . . . Whatever moves, works, or has life, is but an expression or manifestation of Prana . . . Prana is the link between the astral (or auric) and the physical body. When the slender thread-like Prana (the silver cord) is cut off the astral body separates from the physical body. Death takes place. [S. Sivananda, p. 305]

Thus, prana is all around us—"there are currents of [it] running north and south and east and west, and there are also currents of this prana in the body. Those currents are inhaled and exhaled and circulated through the body in our breathing" (Carrington, p. 137). Breathing brings in oxygen and prana. The yoga discipline of pranayana is designed to teach one how to absorb this prana, retain it, and send it through the body, particularly to certain centers to serve vivifying purposes. Many different methods of breath control exist, but all aim at the harmonizing of the inhaling, holding the breath and exhaling, according to a set rhythm.

An interesting approach to prana and vital energy is that of a group of English theosophical doctors and investigators who wrote *Some Unrecognized Factors in Medicine*. The book combines Oriental and yoga approaches to medicine and the investigations of a group of "workers who possess the rare gift of trained psychic capacity," including clairvoyance.

Their starting point is a differentiation of the physical, the vital, and the psychic body.

> The phrase, *the vital body* will be used [they say] as an omnibus term to cover those phenomena of a vital nature which have to do with radiation resulting from chemical changes in cells; with the vital discharges known as nervous impulses; with all that is conveyed to westerners by such words as vitality, animal magnetism or vital energy, and to eastern students by the term "personal prana" . . . Changes in bodily potential and the electrical capacity of the body all fall within the field covered by this term. [*Some Unrecognized Factors in Medicine*, pp. 14–15]

These writers claim that Hindu medical students view solar energy as the source of all forms of energy within our solar system.

> Some western students now claim to have observed that, in sunlight, a subtle change occurs in certain otherwise unattached particles in the air which binds them together into a group carrying a charge of a special growth-stimulating force. This group has been termed the *vitality globule* and is said to be formed only in sunlight. [Reich observed the same phenomenon and called it dancing dots of orgone.] Our observers have also seen it present in large quantities when *ultra-violet light* was used.

> Prana . . . in its behavior . . . is closely allied to elec-
> tricity, with which it has many characteristics in common,
> such as positive and negative reactions, but it is different in
> quality, having an affinity with protoplasm and a life-en-
> hancing action . . . Drawn in as one breathes . . . and dis-
> tributed over the body with incredible rapidity . . . modern
> occult students describe prana as running along the myelin
> sheaths of the nerves and not through the nerve fibres them-
> selves. [*Ibid.*, pp. 31–32]

Furthermore, prana is present in every cell and molecule of
living organisms. It has a special connection with the endocrine glands, and its flow is responsive to the individual's
psychological state and moods. In all these characteristics
prana coincides with Reich's conception of the orgone.

The English theosophical doctors maintain that

> free prana travels over the sheaths of the nerves with great
> rapidity, prana is thrown off from all extremities, all orifices,
> and the pores of the skin. In a healthy body the pranic flow
> from the body constitutes a fine spray of haze all round it.
> There is also a form of *static* vitality closely associated with
> the structure of each cell and organ. The interplay of the
> quick-moving prana, drawn in from the air, with the static
> prana anchored in the cells, constitutes the organism known
> as the vital body. [*Ibid.*, p. 34]

Perhaps what Ravitz and Burr were observing was the static
form of the prana; and what Reich calls "atmospheric orgone"
is the free prana.

The theosophical writers describe the vital body as a purely
physical organism. Its particles are very subtle in structure,
and in their denser manifestations are just visible to ordinary
sight under special conditions. "The currents of energy which
constitute the subtler parts of this 'body' respond to thought,
feeling and will with extraordinary rapidity, and expand and
contract with deep breathing, relaxation and changes of mood.
Hence the vital energies are in constant motion" (*ibid.*, p. 35).
"Those with sensitive fingers can study [feel] its quality and
variations." The vital body extends about one quarter to one
half an inch outside the physical skin. It can be seen fairly
easily by some people and "has the appearance of a rapidly

moving grayish band flowing over the surface of the physical skin. . . . The physical phenomenon most nearly suggesting the appearance of the vital body and its radiations is probably the aurora borealis" (*ibid.*, p. 37). In Chapter Fifteen we will read how Russian scientists who have photographed the aura in recent years suggest that it looks like the aurora borealis.

In health, our theosophical writers maintain,

> the radiations stand out at right angles to the surface of the body, throwing out surplus or used vitality . . . While in disease and fatigue they droop and become tangled. Tangled spots due to local disorder can hold back the circulation of prana . . . and congestion of vitality then takes place locally . . . Pain apparently occurs at the point where tangled streams of vital energy have formed an effective block to vital circulation; the strain of the dammed-up energy . . . bringing about in the . . . physical body a certain nervous pressure which causes pain. Moving pains and spontaneous but transitory itches and skin tension . . . are generally due to shifts in vital pressure, not usually to any actual disease of physical tissues. [*Ibid.*, pp. 37–38]

These notions about energy blockage and resultant physical illness duplicate exactly what Reich claims for blocks to the orgone's flow through the body.

Apparent similarities between prana and the orgone are deepened when these doctors discuss breathing:

> Deep rhythmic breathing draws in a larger quantity of physical prana than shallow or irregular breathing. The supply . . . is greater in direct light and better still in natural or artificial sunlight . . . A few steady breaths affect the whole body . . . strengthening the emanations, raising them if drooping from fatigue and generally brightening and toning up the body. [*Ibid.*, p. 44]

Compare this with the Reichian stress on freed-up breathing outlined earlier.

These English theosophists also discuss the effects of laying on of hands, which they label magnetic healing:

[It] has a direct effect upon the vital body, cleansing, feeding and readjusting it, but the effect of the treatment is very much limited by the personal reactions of the patient, since these also directly influence his vital body . . . The most effective uses of magnetic healing are conditions of shock, both physical and psychological, post-surgical cases, maternity work, etc., [and] in cases needing gradual building up, e.g., convalescence, exhaustion, etc., where a mild steady pressure towards health gradually brings the patient back to his normal condition. [*Ibid.*, pp. 154–55]

These descriptions of healing fit perfectly with many discussions of healing energies having a non-Hindu source—as outlined in Chapter Nine, and with medical applications of orgone. Parallels of interest will also be found in Chapter Sixteen, where I deal with my personal testing of orgone blankets.

The presence of a good supply of prana is the basis for healing by physical contact. So the prana aura which surrounds the physical body not only indicates the health of the person, but "contains physical power and magnetism which may be and is imparted to others" (Panchadosi, p. 18). It is even claimed that small particles of the prana aura are sloughed off by people and can leave an individual's scent in a place. This may help explain why an enclosed place, such as a chapel or church, which has been used for years or centuries for unselfish prayer by truly devout persons will come to have its own "smell," or radiation, and perhaps even function sometimes as a low-level condenser of healing energies. Sensitive people constantly refer to feeling energies upon entering churches or chapels in which over a period of years the faithful have prayed intently.

In yoga philosophy the prana and the vital body are associated with ideas distinguishing other energy sources and powers. Thus, man is thought to have seven bodies from the gross (physical) to the pranic body to the astral, all linked together yet possessing distinctive energy systems. Hindu philosophy also speaks of five tattvas, or ethers—one for each of the senses. These ethers "are thought to be modifications of the Great Breath, [which] flows in five streams—the different tattvas. These [streams] have different shapes" (Carrington, p. 137). Some take the form of triangles, some squares, some loops. They also exhibit different kinds of movements. A few Russian scientists are carefully investigating certain of these, including triangular-shaped forces. This will be discussed in detail in Chapter Fifteen.

In addition, in Hindu and theosophist thinking, the human body has distinct vitality or energy centers, commonly called *chakras*. Located in the prana aura body are seven chakras: (1) at the base of the spine; (2) at the base of the sexual organ; (3) just below the solar plexus; (4) in the heart; (5) in the throat just below the larynx; and (6) between the eyebrows. The location of the seventh is less definite, but it is usually said to reside within the brain (Carrington, p. 149). The six lower chakras may correspond to the six sympathetic plexuses in the body. These chakra points correspond roughly to Reich's division of the body into segments, discussed in connection with bodily armoring in Chapter Two. Reich seems to have "stumbled upon" the same notion of the human organism as has been recognized in India for centuries. In their nature, the chakras can be best described as vortices or "wheels" of invisible energy moving partly in a counterclockwise direction. Disturbances in the chakras lead to ill health.

A special kind of energy, called *kundalini*, is related both to the yoga beliefs in prana and to latent powers residing in these chakras. Some of the ideas about energy flow through the body propounded by clairvoyants or workers in bio-energetics seem to borrow from or duplicate yoga beliefs about kundalini. In addition, some theoreticians of psychic phenomena suspect a connection between kundalini and the more extravagant feats of psychics, such as levitation and psychokinesis.

The kundalini is thought to flow through the body, when activated, in a spiral or coiled fashion. For this reason it has been called a "serpentine" or serpent-fire power. It is supposedly located in the first (muladhara) chakra at the base of the spine, normally in a nonactive state. When aroused it moves through a spinal passage called the "central canal" (Garrison, p. 16).* In theory, kundalini is closely related to

---

* This canal, called *susheurna* in the Hindu lexicon, and two other smaller channels for psychic energy, the *vagrini* and *chitrini*, are invisible to routine procedures of dissection and, hence, denied by Western science.

both sexual and creative energies. Its source is not solar energy, but the magnetic core of the planet (*ibid.*, p. 60). If it is not released in normal sexual or creative expressions, it is supposed to produce mental and physical troubles and perversions. This proposition directly parallels Reich's notion that if the body's bioenergy is dammed up and not expressed healthily, biopathic illness will result.

Various techniques are advocated for awakening the kundalini. One method is to imagine energy flowing down the spine and violently rapping or striking at an imaginary door (the *lotus*) leading into the kundalini's dwelling place. This should be done after the adept has assumed one of various yoga *asanas*, or positions, and completed certain breathing exercises. Most "normal" breathing is very shallow, so "that the current of energy sent downward (unconsciously) to strike at the coiled kundalini . . . does not awaken it. Yoga breathing exercises, on the other hand, send a potent charge of prana coursing toward the kundalini" (*ibid.*, p. 214). Once awakened, this energy is allowed to flow up the spine, animating in turn the other chakras (through mental direction) causing them to spin or glow, and presumably tapping their reservoirs of power.

Carrington, one of the most intelligent and learned of early twentieth-century American psychic researchers, experimented at length with kundalini. His version of how to awaken kundalini is slightly different. (1) He emphasizes: inhaling, holding, and then forcing the breath downwards; (2) mental concentration on the heart chakra, with the image in the mind of a flame existing at that center; (3) mentally projecting the "flame" to the low spine chakra, where the kundalini "sleeps" (the heat generated arouses the kundalini); (4) piercing the opening (door) and proceeding to the second chakra, which is vivified, and so on, till all are finally aroused (Carrington, p. 178).*

It is generally agreed that awakening the kundalini is dangerous and difficult, and involves much mental concentration. The energy aroused seems to want to return to its starting point and rest. The sense of liberation and the psychic powers supposedly conveyed depend on the kundalini's reaching the

---

* For a more detailed description, see Carrington, pp. 168–69 and 176–78.

seventh chakra, in the brain area, and staying there. Carrington suggests that it must be retained there for three days before great psychic gifts become available. Its presence is taken
to be a sign that the individual has attained a merging with
the universal consciousness—that is, a state of harmony of
the forces within the body and those of the external cosmos.

The point of this apparent detour into Eastern thought is to
emphasize that Indian thinking, both Hindu and yoga, on the
body, its energy basis and potentials, is very similar to Reich's.
Already some young Reichians are attempting to integrate the
two approaches; given the interest of many Western youth
in Oriental concepts and philosophy, we can expect interesting
developments from such combinations.

Although much of yoga thinking and practice fits in with
the orgone theory and Reich's view of muscular armoring,
yoga's generally ascetic approach to sex poses a contradiction.
However, there is a little-known branch of Indian thinking,
called Tantra yoga, which not only takes a positive approach
to sexuality but agrees fundamentally with some other
Reichian notions. In fact, Omar Garrison, the author of one
popular study of Tantra yoga, not only quotes Reich favorably
at one point, but reveals a profound and sympathetic understanding of Reich's views on sex and sexual pleasure.

Tantra yoga, while accepting the usual yoga ideas and practices regarding prana, kundalini and breathing strongly affirms the body and takes a sacramental evaluation of sexual
union. More than that, the Tantra philosophy considers the
universal energy to manifest itself "in three ways, as static
inertia, dynamic inertia . . . and a harmonious union of these
opposites" (Garrison, p. xxii). One is immediately reminded
of Reich's basic diagram of the orgone, Ψ , in which the one
energy possesses two opposite characteristics. Again, in Tantra
thinking, the universe or macrocosm in which these energies
flow, is duplicated microcosmically in the human being. In
short, man is supposedly made in the image of the universal
consciousness and should "identify the corresponding centres
in his . . . body with those of the macrocosm" (ibid.). One is
reminded here of Reich's claim, in his study Cosmic Superimposition, that even the galaxies, in their form, exhibit a pattern
of superimposition, ᔕ , like that of sexual union among

living organisms. From this it follows in Tantra yoga that sexual union is fundamentally a religious act as it unites the static (female) and dynamic (male) energies.

It seems that this conception of sex may be pre-Indian, having intimate connections with many early civilizations. The devotees of Ishtar in Babylonia, Isis in Egypt, Aphrodite in Greece, Diana in Rome, may have shared a roughly common conception. In a word, the mystery religions of both East and West probably possessed a somewhat similar outlook on sex.

Thus, the Tantra is a body-focused philosophy, as is Reich's work. Experience in the flesh leads to the divine. For instance, a passage in Rotvasara declares that "he who realizes the truth of the body can then come to know the truth of the universe" (*ibid.*, p. 41). More than that, it is held that the individual is or can be in constant psychic interaction with all of life . . . The Tantra premise is that an exchange of psychic energy is constantly going on between person and person, planet and person, and planet and planet. In his *Cosmic Superimposition*, Reich took a similar stance. Recent research on the planets and organic life is beginning to draw new credence to this view. We shall look at these amazing claims of man's intimate connection with cosmic forces in Chapter Fourteen.

In terms of actual sexual union, the Tantra is aimed at intensification of the senses. An elaborate ritual is to be followed, whereby the couple contemplating coitus maximize the appeal to the senses and deliberately prolong various aspects of physical touching and intimacy, so that the final physical union floods the entire body with a series of blissful sensations. Ritual acts, including the drinking of wine, vivid imagery and ceremonial touching of each other are all designed to emphasize the sacredness of the act. Much self-control is required at early stages of the ritual. (It was probably Tantra yoga that was being satirized near the end of Terry Southern's *Candy*, when the girl and the guru began to lie down together.) The end result is supposedly a sense of liberation, a greater love for the partner as well as a feeling of having participated in a divinely ordered act. The orgasm is held to be life-restoring and life-ordering. In essential, this represents Reich's ideology, although his conception of the physical character of the true orgasm was rigorously defined.

Withal, this brief look at yoga-Hindu beliefs indicates that for centuries millions of their religious adepts have accepted a life style that accords with and utilizes some fundamental Reichian notions. Their approach to unlocking captive energy is different in some respects, but both stress vitalistic concepts, the saliency of deep rhythmic breathing and the right use of the vital creative powers locked inside the organism. In short, the orgone theory may be a Western scientist's rationalistic attempt to capture, unconsciously, certain Eastern insights for Western man.

# Chapter Seven

# THE HUMAN AURA AND OTHER
## AURIC ENERGIES

From Indian philosophies about energy systems, it is a small step to those occult and psychic researchers who are attempting to explicate the human aura and its functioning. Generally ridiculed until recently because of its association with parapsychology, the aura has attracted new scientific interest, particularly in Soviet-bloc countries. Some of these researches throw light in a general way on Reich's orgone theories; others appear to confirm specific aspects of the orgone flow through the body and the role of life energies in the "diagnosis" or treatment of illness and disease.

Historically, there is evidence that some persons in most, if not all, cultures have actually seen subtle rays emanating from the human body. For instance, there is testimony about the aura on the walls of ruins in India, Egypt, Peru and Yucatan. In fact, in the ancient caves of Ellers can be found figures with emanations around the body and head of Buddha and others. And, of course, many pictures of Jesus and the saints depict a halo around the head. Ancient doctors and philosophers have apparently made reference to the aura under many and various names. For example, Hippocrates used the term *enomron*, Pythagoras called it the luminous body, while the cabalists referred to the astral light.

Colonel Albert de Rochas, of France, was one of the early

modern investigators of the aura, which he called the *peri-spirit*.

> He (himself) saw a shadowy outline around the body of a
> sensitive; if he touched this layer—when the sensitive's back
> was turned—the patient felt it, sometimes as a burning sensa-
> tion. If a glass of water was passed through this zone of ex-
> teriorization . . . and then taken away the water remained
> sensitive and could be distinguished from ordinary water by
> taste. De Rochas . . . sought for the substances which were
> the best for storing this exteriorized sensitiveness (or energy)
> and found that liquids, viscous substances, gelatin, wax, cotton,
> wood and velvet were best. [Poinsot, p. 451]

These compare closely with Reich's ideas on which substances
hold orgone.

Other French scientists from the late-nineteenth century, including Baraduc and Boirac (see Chapter Four) believed their
researches established the existence of an aura. In this century
in Britain, Dr. Richardson claimed evidence for a "nervous
ether," which he defined as a finely diffused form of matter
pervading the whole body and surrounding it in an enveloping
mantle (Westlake, p. 10).

The first of a number of attempts to make the aura visible
to anyone was initiated by Dr. Walter Kilner, of London, who
described his experiments in 1911 in *The Human Atmosphere,
or The Aura Made Visible by the Aid of Chemical Screens.*\*
Revised and republished in 1920 the book "was sympatheti-
cally reviewed in *The Medical Times* (Feb. 1921) and in the
*Scientific American* (March, 1922) but . . . had a rather limited
circulation" (Bagnall, p. 11).

Kilner's "findings" seemed to demonstrate a physical emanation around the whole body:

> If someone was placed against a dark background in twilight,
> a slightly luminous mist, oval in form, was seen round his

---

\* Kilner, a reputable member of the London Royal College of Surgeons
and a man with no previous interest in the psychic, based his book on
four years of experiments in London hospitals. He made a device of
colored screens, using a little-known chemical called dicyanine, which
gives in alcohol solution, a beautiful violet blue. By looking through
these screens, Kilner felt that anyone could see at least a little of the
auric emanations.

body. This had three distinct zones. The first was a dark edging, half a centimetre wide, surrounding the body; this was the "etheric double." Outside this was the interior aura, dense and streaked perpendicularly to the body; this was from three to eight centimetres in width. Finally came the exterior aura, which had no definite contour. [Sudre, pp. 113–14]

The visual phenomenon varied substantially with the age, sex, mental ability and health of the subject. The color was often bluish. If various gases, such as chlorine, were liberated in the vicinity of the subject, the aura would take on different colors. When the pole of a magnet was brought near, a ray seemed to be formed temporarily which joined the pole to the nearest or most angular part of the body. A charge of static electricity momentarily dispersed the aura, but led to its intensification afterward. These and similar tests convinced Kilner that he was observing an objective emanation. He felt that 95 percent of people could see the aura through his screens. He even went so far as to build a system of medical diagnosis upon what he observed as to the color, texture, volume and general appearance of patients' auras.

Commenting on what one sees through the dicyanine screens, H. Boddington writes:

In certain diseases there will be void spaces which indicate diseased organs . . . The aura is not usually seen covering the entire body at the same moment, because of its exceeding delicacy and transparency. Nor will the colours be seen in sharply defined patches. One screen emphasizes one aspect but cuts out another colour. Disease shows itself by irregular patches, or as spots in special localities. The aura extends about eighteen inches from the body and is more condensed the closer it approaches the body. The densest section lies against the skin and often eludes sight altogether, but is revealed by the secondary screen which, in turn, cuts off the appearance of the outer aura . . . During trance states Kilner found the aura almost disappeared, [and] at death, it almost entirely leaves the body . . . In short, apparently there is a close link between the aura and the activity of the *conscious* mind. Numerous observations including its diminution at the approach of a fainting spell bear out this observation. [Boddington, p. 8]

Some efforts to duplicate Kilner's findings have had negative results. A. Hofmann, a German investigator,

> made a series of colour filters and studied their effect on vision. He concluded from his rigorously detailed experiments that the aura was a completely subjective impression caused by fatigue of the retina and by a difference of accommodation for different colours. A decisive proof of this was given when, after retinal "sensitization," an aura could be seen round a plaster bust! [Sudre, pp. 215–16]

Sudre, an experienced researcher in parapsychology, tells of spending hours staring through a dicyanine screen and seeing nothing. It seems true that the dicyanine screens were useful to only a small percentage of observers, and that Kilner's success probably was attributable to his unwitting possession of clairvoyant powers.

Some present-day clairvoyants who seem to see auras furnish similar pictures of what they perceive. They include such internationally known figures as Edgar Cayce, Eileen Garrett and Ambrose Worrall, the healer. Just before he died Cayce, with the help of his biographer-friend, Thomas Sugrue, wrote a small booklet entitled *Auras*. While empirical scientists may not be impressed, the enormous reputation that Cayce has gained in recent years for special insights makes his views significant.

He begins: "I do not think of people except in connection with their auras; I see them change in my friends and loved ones as time goes by—sickness, dejection, love, fulfillment— these are all reflected in the aura" (Cayce, p. 5). He found that clairvoyants, as well as writers on the psychic, are generally in agreement with him as to the aura colors and their meaning. People tend subconsciously, he claimed, to choose clothes that fit with their aura. When they don't, we subconsciously notice how the aura and the clothing clash and say, "Why does she wear that color?"

A fascinating part of the booklet refers to Cayce's interpretations of auric colors. Red indicates force, vigor and energy. Dark red indicates high temper and is a symbol of nervous turmoil (*ibid.*, p. 11). Orange denotes thoughtfulness and consideration, but the shade is important—a golden orange shows self-control, whereas brownish orange can indicate a lack of ambition or an "I don't care" attitude. Yellow stands for health and well-being. People with yellow in their aura are happy, friendly and helpful, but if the yellow is ruddy, they are timid. Green is the color of healing, commonly found among doctors and nurses. But a lemony green, with a lot of yellow, indicates deceitfulness. Blue stands for contemplation, prayer; the deeper shades mark people of great spiritual vigor. Pale blue shows little depth but a struggle toward maturity. Indigo and violet are indicators of seekers, those looking for a cause or a religious experience. The perfect color, all clairvoyants agree, is white, which symbolizes great purity.

Cayce emphasized the importance of the intensity of the colors, their distribution and positioning:

> The aura emanates from the whole body, but usually it is most heavy and most easily seen around the shoulders and head, probably because of the many glandular and nervous centers located in those parts of the body. The dark shades generally denote more application, more will power, more spirit. The basic colour changes as the person develops or retards, but the lighter shades and the pastels blend and shift more rapidly as the temperament expresses itself. The mind, builder of the soul, is the essential governing factor in the aura; but food, environment, and other conditions have their effect. [*Ibid.*, p. 16]

Professor Eugene Osty, in his *Supernormal Aspects of Energy and Matter*, has reported on a significant series of experiments that he conducted with an Austrian psychic, Rudi Schneider, whose gifts included incredible powers of psychokinesis. In one experiment, which followed the strictest scientific precautions and included the use of an ultraviolet camera and infrared beams, it was established that the "substance" leaving Schneider's body exhibited rhythmic movements coinciding with his cycles of respiration. This "substance" was able to move heavy objects several or more feet away from

Schneider, who was strapped in a chair. It was intriguing that Schneider's respiration cycles went as high as 300 per minute, whereas a normal cycle is only 15 per minute. (The pulse rate showed insignificant variations from the norm, and blood, when examined, indicated no important variations.) Osty theorized that the rapid breathing was the basis for the severe neuromuscular labor required to exteriorize his force field (aura) and so move the objects. Moreover, Schneider could exercise control over the force by his thought processes to the extent that he would announce in advance the actions, density and quantity of the exteriorized psychic "materials," and could at times cause it to enter into the experimenter's infrared "cage" on demand. Schneider usually produced a "substance" sufficiently condensed to be opaque to the infrared light. However, Osty once saw with his naked eye a sort of dense fog flow toward the table, which moved when the substance reached it.

These experiments of aura exteriorizing take on fresh interest when compared with tests performed in the sixties by Soviet scientists upon their most gifted demonstrator of psychokinesis, Mrs. Nelye Mikhailova. Her amazing psychic abilities were captured on film and shown at the 1968 Conference on Parapsychology in Moscow. Conducting a test on Mrs. Mikhailova during a filmed psychokinetic performance, the Russians found her heartbeat moved up to 240 beats a minute, about four times the average. Other machines wired to the psychic, as she moved small objects by psychic force, showed (a) that the EEG's were exceptionally active, and (b) that the force field around her body—much bigger than average— began to pulsate. "Brain and heart pulsed in rhythm with these vibrations in her force field" (Ostrander and Schroeder, p. 75). The researcher Genady Sergeyev theorized that such pulsations caused the objects to be moved as if magnetized—i.e., they moved from or toward the psychic, as if she were a magnet.

It is an educated guess that the aura of a psychic is different in degree, but not in kind, from those of ordinary people. One of Osty's sensitives describes the aura as "an indefinable fluid that radiates from the whole person . . . composed of several elements . . . light, heat, vibration, electric or magnetic currents . . . But these elements never present themselves equally.

No two [auras] . . . present the same appearance" (Johnson, p. 187). Physical mediumship, according to many sources, arises from some peculiarity of state or structure of the medium's auric or etheric body. It may occur at puberty when the aura often undergoes a great change. Some parapsychologists believe that the etheric body is the source of the dammed-up energy which is thought to be the cause of poltergeist phenomena.

Eileen Garrett, one of the twentieth century's most noted mediums, often wrote of the aura and its energies. A highly successful business executive and founder of the Parapsychology Foundation,* Mrs. Garrett, wrote in her book *Awareness:* "Throughout my whole life, I have been aware of the fact that everyone possesses a second body—a double. The double is a distinct fact in Eastern and theosophical teaching and as such it is said to be an energy body." She adds, "For myself I know that the vitality of the human organism is a synthesis of energies which are being constantly received, transformed, fused and expressed by the organism. I've always seen plants, animals and persons encircled by a misty surround" (Garrett, p. 106). Then, specifically, she states, "The double is the medium of telepathic and clairvoyant projection" (Ostrander and Schroeder, p. 211).

Perhaps the most gifted clairvoyant healer in North America since the forties was Ambrose Worrall, of Baltimore. Mrs. Olga Worrall, his wife, is also a clairvoyant and has teamed up with her husband in all his healing activities. Mr. Worrall, in addition to being a consultant to the large Baltimore corporation Glenn L. Martin, has performed thousands of healings over the past forty years. Several books attest to his wisdom and extraordinary healing powers. In *The Gift of Healing,* he said: "We are surrounded by a healing power, just as we are surrounded by magnetic fields, electricity, air, ether and all the other forces around us that we do not see but are able to measure and study in various ways. The need for scientific knowledge about this field force of healing is obvious" (p. 173).

---

* A leading intellectual organization in the Western world dedicated to research and to study of the psychic.

Dr. W. W. Coblenz, an outstanding American physicist, expressed his findings in 1954 in *Man's Place in a Superphysical World*. Coblenz attempted an intensive discussion and an examination of researches into the psychic. In the Introduction he wrote: "I join with Alexis Carrel, Charles Richet, Hamlin Garland and Alfred Still in a plea for less dogmatism and more intelligent inquiry into this almost unknown aspect of the mind" (Coblenz, p. vii). He examined numerous manifestations of the psychic, and he apparently confirmed scientifically many of the basic ideas of parapsychology. With respect to the aura, he investigated the effect of magnets on the human double. Utilizing a reputable sensitive who saw an aura around a magnet, but not around an unmagnetized soft-iron bar, he reported: "The magnets had an aura which operated to interfere with the tests on the auric light emanating from my body" (*ibid.*, p. 9). Here is more support for the repeated claim that magnets affect the human mind-body.

One of the most mature and reasonable approaches to the concept of life energy and the aura is that of J. Cecil Maby, a British scientist involved in physics and radiesthesia. The following presentation of his viewpoint, although not comprehensive in that it is based solely upon a few articles he wrote for *Radio Perception,* is enough to illustrate his contribution to the study of the aura. A fuller development of his concepts may be found in his major works, *The Physics of the Divining Rod,* written with T. B. Franklin (London: G. Bell and Sons, 1939), and *The Physical Properties of Radiesthesia* (Birmingham, England, 1966).

Maby, a recognized scientific authority on water divining, whose divining services were engaged by the British government during the Second World War, approaches the field of radiesthesia and the psychic in a very cautious and critical fashion. He emphasizes the necessity of assuming, until proved otherwise, that all the unknown energies in these fields are physical and not vitalistic. In this, and in his whole approach to radiesthesia, he follows in the tradition of European investigators like Frederick Pohl, Budgett, Trinder, Applegate and Solco Tromp. He is conversant with the literature on vital energy, the thinking of Mesmer, Charcot, Reichenbach, Boyd and Eeman, as well as the electronic and radionic medical machines of Albert Abrams, W. E. Boyd, Ruth Drown and George de la Warr.

Maby defines the radiations from the human organism (which he notes are stronger when the person is emotionally aroused) as psychoradiant or electromagnetic rays. With the aid of an instrument called a radio electrometer,* which he perfected, he picks up this energy release, labeling it a short-wave Hertzian effect. Similar machines, he points out, such as Cazzamalli and Boyd's Emanometer and Wigelsworth's Pathometer, have been used for the same purpose. Cazzamalli, an Italian neurologist, developed an apparatus in the late twenties which led to the claim that centimeter-long electromagnetic waves move in the space around the human head. These and longer waves particularly were measurable by Cazzamalli during heightened mental activity; other investigators were not always able, however, to duplicate his results. Boyd's Emanometer and Wigelsworth's Pathometer have received less scientific interest.

Maby states his findings as follows:

> In the case of the human psychoradiant phenomena, I find that an imaginative person can project energy capable of affecting either the radio electrometer or a radioactive preparation in conjunction with a Geiger counter, without any bodily movement or even invisible muscular tension, and that over distances of many yards; also by visual concentration alone—which creates a beamed effect of opposite polarity from the Right and Left eyes, as also from the R and L hands . . . And the hypothetical rays can be reflected, refracted, absorbed, polarized, etc., in a way that suggests electromagnetic waves, that almost certainly lie in the micro-Hertzian region of the spectrum, between the ultrashort wireless waves and the infrared—as other investigators have also surmised. Various screening tests indicate that they cannot be visible light rays, ultraviolet or even heat rays, and nearly all reaction is stopped by quite thick, but continuous metal screening, also by 1/2 mm. mesh wire gauze. [*Proceedings of British Congress on Radiesthesia, 1950,* p. 86]

---

* It is described in *Radio Perception*, Vol. VI, No. 49 (1945). Maby has made a color movie which illustrates the operation of his electrometer.

Maby cautions intelligently against lumping little-known forces under any one heading such as "vital energy." He emphasizes that

> dowsers themselves respond to a wide selection of rays and fields (including at least Hertzian, blue-violet, ultra-violet and gamma rays), and that human detectors are hard to use in the laboratory. Add "psychic" faculties, and auto-suggestion, electro-magnetic induction, aerial ionization, sonics and ultra-sonics, and even static potential effects, and one is properly in the soup! Instruments are at least less catholic in behaviour and more controllable . . . Excessive sensitivity opens many booby-traps, and multiple factors keep cropping up that are extremely hard to sort out. [*Ibid.*, p. 87]

Such skepticism concerning the specific reliability of human detectors of energy fields is eminently sensible.

One of the most interesting discussions of the scientific basis to dowsing is carried in Ostrander and Schroeder's *Psychic Discoveries Behind the Iron Curtain*. They relate how dowsing has been accepted in the highest Russian scientific circles—for example, the Geology Department of the prestigious Moscow State University—and how it is being widely used to discover valuable ore supplies. Thus at a two-day conference held in Moscow in 1968, "delegates reported, for instance, that the 'biophysical effect method' was being used in the Yakut Republic in northern Siberia; in engineering geology in Lithuania; in the search for water in the desert; in prospecting for ore deposits in Central Asian U.S.S.R.; and was being extensively studied in field and laboratory by Leningrad and Moscow geology institutes" (Ostrander and Schroeder, p. 193). (Dowsing has been renamed Biophysical Effects Method, B.P.E., in Russia.)

Some of the more pertinent quotes from their discussion, in terms of orgone theory are:

> Dowsers find it hard to get a reaction during lightning storms. Apparently changing weather and geophysical conditions cause the force from the minerals to reflect like rays of light at different angles . . . Russian dowsing "operators" also took to the fields in trucks, cars, and buses. . . . "The metallic body

of the bus or car didn't influence the dowsing reaction . . .
Inside or out of the car, the dowser still got the identical
response. It means that this unknown energy is *not* electrical,
because the metal body of the car would screen out electrical
energy and insulate the dowser from the ionized fields of the
earth" . . . When a seasoned dowser touched the hand of a
nonoperator during a test, the divining rod in the nondowser's
hands suddenly came to life. The sensitivity to [dowsing]
forces . . . could somehow be transferred from person to
person . . . Professor J. Walther, of Halle, West Germany,
found dowsers showed higher blood pressure and pulse rate
above a dowsing zone. Dr. S. Tromp, a Dutch geologist researching dowsing for UNESCO, reported, in the Winter 1968
*International Journal of Parapsychology*, that the body's reaction to water or minerals in the earth can be clearly registered
with an electrocardiograph. [Ostrander and Schroeder, pp.
189, 190, 191]

When in the late forties Reich gave some attention to the
phenomenon of dowsing, he accepted it as a natural expected
activity of the sensitive or less-armored human organism. As
might be expected, he held that it was orgone energy that
moved up from underground water sources to course through
the dowser's body and pull down the willow wand. While
it may be that orgone is involved, many will, in the absence of
definite experiments, agree with Maby that one should be
prepared to see the role of waves and energies of a less
ultimate character. Yet Reich may be on sound ground when
he suggests that to be a good dowser one needs to have an
open, genital-type character structure, so that the energy can
flow readily through the body. There is impressionistic evidence for this. In the one experiment in dowsing in which I
personally was a participant it was easily seen that inhibited
personalities had almost no success, while the more open,
outgoing, lively persons displayed some noticeable capacity
for dowsing.*

The most recent discussion of the aura to come out of

---

* This was carried out at Wainwright House, Rye, New York, at the
Third Seminar on Spiritual Healing, sponsored by the Laymen's Movement for a Christian World.

Britain is a small book by Oscar Bagnall entitled *The Origin and Properties of the Human Aura* (1970). Bagnall has had some education and training in biology, chemistry and physiology, and he approaches the topic in an unemotional and scientific fashion. He was originally interested in the work of Dr. Kilner and has continued in that tradition. His approach strictly avoids any mention of psychic gifts, the occult or the esoteric.

Bagnall's contribution is twofold: first, he makes use of a new dye and new apparatus, to facilitate everyman's viewing of the aura; and second, he tries to explain its makeup in terms of known (invisible) energies. Instead of Kilner's dicyanine, he uses a collateral chemical called *pinacyanol* which he inserts into a glass trough arrangement. By looking through this, the eyes soon become sensitized and eventually a thin inner and a wider outer aura can be seen. He thinks that the dye affects night-seeing nerves of the retina—called retinal rods—enabling one to see shorter wave lengths than one could see under normal conditions. The wave lengths that make up the auric emanations, he believes, are largely from the ultraviolet part of the light spectrum. It is Bagnall's opinion that the pinacyanol, as it sensitizes the eyes, conveys the effect of wearing a convex lens. Thus, it opens up a section of the ultraviolet spectrum not usually available to human beings. (Animals like cats and owls are apparently not so handicapped.) He claims that the aura rays are presumably in the 400–300 $\mu\mu$ range.

Bagnall describes the construction of his viewing apparatus in the following way: "Two pieces of flat (thin) glass are cemented together so as to make a narrow trough, the sides being some 3 to 5 *mm.* apart. The exact distance apart will . . . depend on the strength of the solution [it] . . . is to contain" (Bagnall, p. 48). The pinacyanol is dissolved in alcohol, the "screen" (trough) is filled with the solution and then sealed up to avoid evaporation. For better viewing, the screen or lenses can be mounted on collapsible goggles and fastened behind the head. No light must be allowed to reach the eye except through the screen. To see the aura, the eyes must first be sensitized by gazing at the sky for a couple of minutes. Then the surrounding foliage or other living objects are examined. Next, as the fingers are drawn apart, the rays that shoot out from their tips may be seen (*ibid.*, p. 63).

Bagnall cautions the beginner not to expect too much at once. "One may not be able to see anything but a blur for some time . . . The dye's power is cumulative and gradual . . . If you see nothing at first don't strain your eyes by staring too hard . . . Do not use the screens too long at one sitting" (*ibid.*, p. 50).

A brief review of what Bagnall claims can be seen is fascinating. Not at once, but gradually, an inside aura can be detected—extending up to about three inches from the skin—and an outer haze or mist of a blue-gray or lavender color extending six to eight inches further out. "Probably the aura when first seen will appear as a lighter-purple vapor around the darker-purple skin. Later the differentiation . . . [of the two auras] will be picked out" (*ibid.*, p. 56). The inner aura is striated, its rays normally running parallel to one another and at right angles to the body. The mist or outer aura appears to be never quite still. To see it the observer must not stare directly at it, but slightly to one side of it. A backdrop of red or black (cloth) will help. Occasionally, rays of a somewhat brighter light will be seen extending from the inner aura into and even outside the outer one. They may run toward an object like the pole of a bar magnet. In women, the outer aura is somewhat bigger and more oval than in men; also, it is often larger during menstruation and during the later months of pregnancy. The widening of the haze emitted by the mammary glands is a reliable indicator of the early stages of pregnancy. In general, the outer haze is affected by nervous disorders and by changes due to sexual and, to some extent, hormonal action.

A variety of tests have led Bagnall to some provocative conclusions as to the constitution of the two auras. Thus, while the inside aura will "run to a magnet," the outer is impervious to magnetism. Nor does the inside aura show any polarity. An electric charge—e.g., by a Wimshurst machine—will affect (change the size of) the inside aura, but not the outer. Since the inner aura extends out further from pointed parts of the body like the fingers and the elbows—when hands are on the hips—it behaves like a charged conductor, which has a wider

field around points than around flat surfaces. Thus Bagnall
implies the inner aura is made up of electromagnetic particles;
the outer, he believes, is in the ultraviolet radiation band.

In spite of a strong skeptical bent, he says that "a highly
charged person may be able to impart (by laying on of hands)
a slight, though nonetheless beneficial, current to another body
with a lower potential" (*ibid.*, p. 94). Confirmation of this
statement comes from many psychics—for example, Margery
Bazett, who said that "While observing the British healer
Parish lay hands on a patient, she saw 'streams of blue light
emanate from his fingers and hands as he passed them over
the affected parts' " (Bazett, p. 85). Thus, Bagnall, through his
aura studies, independently provides some rational explanation
for psychic healing. And while he stresses the natural *electric*
and *ultraviolet* aspects of the aura, he concludes: "Clearly
something more than mere ultraviolet rays, or machine-made
electricity is emitted by the living body, in the form of the
aura" (Bagnall, p. 96). In short, this cautious researcher hints
at some vitalistic or orgone-type theory to account for what
he has observed.

Writing on "The Energy Field of Man" in the British neo-
Reichian quarterly, *Energy and Character*, psychiatrist John
Pierrakos adds still new insights to our understanding of the
aura. He began to study this in the early fifties when introduced to orgonomy and the accumulator. He emphasizes: "the
field phenomena are related to the energy metabolism of the
body, its production of heat, emotional excitement, rate and
quality of breathing, activity and rest. They are also affected
by atmospheric conditions, relative humidity, polarity of
charges in the air and many other unknown factors" (Pierrakos, "The Energy Field of Man," p. 65).

Like other investigators,* Pierrakos describes three fields
around the human body. He admits to great difficulties in producing an exact description, partly because of their characteristic pulsations. Apparently the outer two layers appear for
about one quarter of a second, vanish in about one fifth to

---

* For instance, clairvoyant Margery Bazett writes: "I am inclined to
think from personal experience that there are three [zones of auric
emanations] . . . one is . . . the physical, the other two being the astral and the etheric" (Bazett, p. 79).

one eighth of a second and then reappear after a pause of one to three seconds. "This process is repeated 15 to 25 times a minute in the average resting person" (*ibid.*, p. 66). Within the second or intermediate layer, three to four inches thick (this apparently corresponds to Bagnall's inner aura), there are primarily three patterns of movement—wavelike, corpuscular and linear. This last, which takes the form of white or yellow rays, may extend from the intermediate layer several feet outward. It can change its pattern and outline with every new pulsation of the organism. The third or outer layer is almost transparent and can in an open space extend considerably beyond its usual width of six to eight inches. Its predominant movement is spiral or vortical.

Within the total auric body, says Pierrakos, the direction of movement of the energies, when the organism is standing and facing the observer, is somewhat like an elongated figure eight. In one phase, energy moves up from the ground, on the inner side of the legs and thighs, up the trunk and the *outer* side of the hands, forearms and arms; "the two main streams meet and travel upwards towards the neck and over the head" (*ibid.*, p. 68). Simultaneously in a second phase, energy moves toward the ground, on the inside of the trunk. From another perspective, the movement is both along the longitudinal axis and the body's diameter. The longitudinal keeps the body erect, while the transverse has to do with expansion and contraction. In sum, "there are spiral movements fusing the two halves of the body and the two halves of each organ" (*ibid.*, p. 69).

Pierrakos emphasizes the complexity of the energy picture. He says: "The energy field is actually a reflection of the multitude of energies moving and expanding in all directions within the living body" (*ibid.*, p. 69). Specific observations he has made are of interest—for example, in play or excitement the pulsation of the field can double to 40–60 pulses a minute. In great excitement, the layers will double their width. Swimming can double the pulsation rate, deepen the field and increase the brilliancy of all layers. In short, swimming is energizing. Reich emphasized its therapeutic and energizing capacity. In the full sexual act, the field also widens greatly and the energy flow increases enormously. Where a lecturer excites his audience, the individual pulsation rates speed up, and an integrated field joining the listeners' heads to him, pulsates at a faster-than-usual rate. The opposite occurs where an audience is bored.

Character structure is correlated to auric form. Rigid character structures may have a good general flow, but reveal "pathologies" tied to specific blockings of the organism. Specific character types display specific changes in the field. For instance, people with an oral-type character show a weak, narrow field. Many people seem to show a disturbance in the small of the back and around the top of the ilium. It appears as a granulation of the intermediate layer. In those who are physically debilitated, the pulsation rate falls below 15 per minute, the layers are reduced to perhaps half their usual width and the color goes dull and gray. In schizophrenics the coloring tends to be yellow-brown, the pulsation rate below 15, the energy movement twisted especially around the back, and there are numerous interruptions in the smooth flow of the energy.

In a second article in the same quarterly, Dr. Pierrakos describes the energy field around plants and crystals. He claims that he sees plants surrounded by "a radiant luminous field which pulsates rhythmically from ten to thirty times a minute" (Pierrakos, "The Energy Field of Plants and Crystals," p. 22). He likens it to a pyrotechnic of many hues, beams and fireballs. The field has two layers: one is one sixth to one eighth of an inch in size and over-all light-blue or gray in color, and the second is one half to one inch thick and including various multicolor radial movements. Fireballs may shoot out from this outer layer. Colors are specific to the plant and its growth stage. The plant's orientation to the geographic cardinal points seems to affect its pulsation rate.

Trees have a pulsatory energy field which also varies with the species. Evergreens, for instance, pulsate eighteen to twenty-two times a minute, almost twice as fast as the deciduous. Observations of large wooded areas indicate that the "individual field of each tree merges with its companion . . . and the forest as a whole produces a pulsatory movement upwards from the roots to the leaves" (*ibid.*, p. 23). In a forest the normal inner field of three to four feet may be elongated upwards for hundreds of feet. As with all plants, the pulsation includes a phase when the energy of the surrounding atmosphere streams into the top of the trees.

As noted in Chapter Five, Pierrakos has conducted some simple experiments on the interaction of plants like chrysanthemum with the auras of patients seized by depressed or hostile emotions. He noticed the plants' field would shrink, and the pulsations slow down significantly. Various observations convinced him that "there is a constant interaction between the energy field of plants and the energy field of the human organism" (*ibid.*). In this connection it is interesting that the famed British healer, Harry Edwards, has advised people who are low in vitality, to stand close to a large healthy tree and let its energies replenish one's physical powers.

The widely acclaimed discovery of Mr. Cleve Backster,* of New York, that plants have feelings substantiates Dr. Pierrakos' claims. Mr. Backster, an acknowledged expert in lie-detection technology, has demonstrated in the past several years that plants like the rhododendron respond to threats to their life with anxiety states—recorded electronically—and also can become simpatico to specific humans and mirror their emotions. He has demonstrated this phenomenon on TV and his work has been discussed recently in leading American periodicals. Scientists and engineers especially have been captivated by his rigorously controlled experiments. All together, the work of investigators like Pierrakos and Backster adds greater probability to Reich's claim that orgone energy pulsates through all living systems and that all are interdependent, existing in a kind of energy ocean.

---

* See article by Richard Martin, "Be Kind to Plants—Or You Could Cause a Violet to Shrink," in *The Wall Street Journal*, Wednesday, February 2, 1972.

*Chapter Eight*

# ENERGY THEORIES FROM THE FAR EAST

For centuries widespread acceptance and medical use of a life-energy concept has been basic to Far Eastern societies, including China, Japan, Korea and Hawaii. The first three made use of it in the practice of acupuncture, a form of medicine that dominated China and Korea, and was important in Japan until the nineteeth century. In spite of the growing influence of Western medicine, in recent decades a revival of acupuncture has occurred. Recently it has spread to the West, so that today acupuncture is used in some hospitals in France and Germany, and taught in four universities in Russia, and gradually being introduced into Britain and North America.

On the island of Hawaii native doctors still function in terms of a life-energy concept, named *mana*,\* which is used both in healing the sick and in performing supernormal wonders akin to those of Western psychics. Max F. Long brought this concept to the attention of the Western world in the early fifties, forming a Huna Research Institute in the United States to experiment with the ideas and practices of the *kahuna* doctors and their skillful use of mana energy. These two energy theories will be examined in turn.

---

\* Max Long seems to use this term *mana*, common in anthropological literature, in a slightly different and more concrete form than most other scholars.

What makes acupuncture more respectable in the West is the new and excited interest in this medical system and its energy theory among Russian scientists. In their study of *Psychic Phenomena Behind the Iron Curtain,* S. Ostrander and L. Schroeder note that acupuncture is being studied at such leading Russian institutes as the Gorky Medical Institute and the Kirov Institute of War Medicine in Leningrad and that Russian scientists have accumulated a great deal of experimental data on how acupuncture works. They say: "Acupuncture has been amped-up, modernized, combined with EEG's, electrocardiograms" (Ostrander and Schroeder, p. 226) and examined by a special kind of photography. Moreover, a Russian doctor and engineer have invented an electronic device called a *tobiscope,* which locates the acupuncture points to within a tenth of a millimeter. This device was displayed at Expo '67 in Montreal as evidence of yet another Soviet first in technology.

What is acupuncture and what is its energy theory? Fundamentally, the term refers to a therapeutic method in which the skin is superficially pricked or "punctured" at specific points by very thin needles, with a view to correcting what are considered to be pathological energy flows within the body. The practice has been traced back some five thousand years, when certain Chinese observers noted that an injury or wound in certain parts of the body—often caused by weapons made of sharp flints or stones—had a curative effect on specific illnesses from which wounded persons had long suffered. For instance, a sharp wound on a certain part of the shoulder might cure or minimize arthritis and its pains. Continued observations showed that the size of the wound or cut was unimportant; what mattered was the exact spot on the skin where it was made. Eventually it appeared that these spots were quite small, roughly the size of the head of an ordinary pin. In time, Chinese medical men went from using pointed fishbones to sharp needles to prick the skin at such points. By extensive examination they found some seven or eight hundred of these critical acupuncture points on the human body. Kirlian photography in Russian labs has in recent years seemed to verify many of these acupuncture points. "After studying the Kirlian process, Dr. Gaikin (a Leningrad surgeon) and engineer Vladislav Mikolevsky, also of Leningrad, invented an electronic device that pinpoints the acupuncture points . . . it is called the tobiscope" (*ibid.*, p. 225). German engineers have also produced a machine which plots an electrical energy flow at these points and it is now being used by doctors and chiropractors there in the diagnosis of many diseases.

An article published as recently as December 6, 1971, in the *Journal of the American Medical Association*, by Dr. E. Grey Dimond, provides some fresh light on the use of acupuncture in modern China. This article is based on a journey undertaken by the author in September 1971, in company with Dr. Paul Dudley White. Dr. Dimond notes that the simple placement of needles "at strategic points rigidly defined by texts with 2,000 years' authority" (Dimond, p. 1159) has been widely replaced by what is called new acupuncture. This involves "new points of insertion and on occasion deeper placement, but also constant manipulation of the needle" (*ibid.*, p. 1160). It "can be described as a rapid up-and-down traverse of the needle over a distance of approximately one-half inch simultaneous with a rapid to-and-fro twirling of the needle by the thumb and fingers" (*ibid*). He adds that there is "also some therapeutic rationale to some phases of acupuncture" (*ibid.*).

Later in this article, we are told of how Western-trained Chinese doctors regularly used acupuncture for anesthesia during operations. The two American doctors witnessed several serious operations performed under acupuncture anesthesia and commented on the patient's absence of pain. In these operations a biphasic pulse generator delivering 9 volts, from 120 to 180 cycles per minute was attached to the needle. Since 1958 one Peking hospital had performed 4,900 operations under acupuncture anesthesia. Two theories are currently advanced, that the needles affect either a neural pathway or the meridians of traditional acupuncture. Two interesting observations made by a Western-trained Dr. Chow were that "following acupuncture anesthesia, there was some leukocytosis and a faster circulation time" (*ibid.*, p. 1563). Questioned by Dr. Dimond, he said that "he, as well as practically every Western-trained physician that he knew, had been thoroughly skeptical of acupuncture anesthesia and had thought that it was essentially a hoax. It was only after repeated personal clinical experiences that he became convinced. Gradually, through guidance from his traditional associates, he had learned how to carry out the procedure, where to insert the needles, and what its limitations were. He believes that it is simply a method that Western physicians must now recognize and that the results of basic research explaining the rationale would be available soon" (*ibid.*).

What interests us most, of course, is the theory behind these acupuncture "probes" and what kinds of disease they appear to help cure. Simply put, the theory is that the human organism is responsive to the total external environment—that is, that a man is linked to a cosmic vital energy. If there is a change in the energy envelope around man, he will be affected by it; the vital energies in the body resonate to these changes and they in turn affect the physical body. This vital energy, for instance, is affected by such things as the seasons, the cycles of the moon, tides, thunderstorms, strong winds and even levels of noise. "The State University of Kazakhstan affirmed that the bioplasmic body is affected by changes in the atmosphere just as acupuncture theory predicted" (*ibid.,* p. 225). This fits in with findings reported by Burr and Ravitz in Chapter Five. Further evidence on cosmic energies and their effects on the human organism are discussed in Chapter Fourteen.

The details of acupuncture's vital-force theory also bear out much of what Reich and other vitalists have contended. The vital energy (*tch'i*) is polarized into a positive phase *yang* and a negative one called *yin*. Yang represents activity, virility, the male, the sun, the day, heat and summertime. Yin stands for the opposite: rest, the female, night, cold, the moon and winter. Besides being subject to the rhythm of activity in the day and rest at night, man is also influenced by an inner rhythm. Indeed, acupuncturists emphasize the rhythm of the heart, contracting and expanding, the in and out of respiration, the contraction and relaxation (expansion) of the bladder, the peristalsis of the digestive tube. As Louis Moss says in *Acupuncture and You* (p. 21), "all follow the 'cosmic' pulsation." Here is fundamental agreement with Reich's emphasis on the essence of life being pulsation, contraction and expansion. Again the Chinese medical men stress that life depends on the *balanced* and *harmonious* circulation of *tch'i* —i.e., life energy. Here again is agreement with an orgone postulate. In addition, "the ancient Chinese treatises affirm 'the blood circulates following the energy.' If the energy circulates, the blood circulates; if the energy is trapped, the blood stops" (Moss, p. 21). Thus, if tense musculature dams up the flow of energy, it retards the blood's circulation, and disease will eventually set in.

Acupuncture is especially useful, it turns out, in treating functional complaints like neuralgia, gastric complaints, rheumatism and arthritis. In justifying the usefulness of acupuncture Moss quotes Hans Selye on the serious effects stress and resultant muscle rigidity can have on the health of organs; he might as easily have quoted Reich, but he probably did not know of his work. In short, acupuncturists see Selye as on their side, and they would equally strongly endorse Reich's muscular-armoring and orgone concepts.

Specifically, acupuncture sees the vital energy flowing through the body along lines called *meridians,* which connect organs with the points on the skin surface. For instance, a point on the inside of the little finger on the right hand connects through energy pathways to the heart. Furthermore, the nerves pricked by the needles belong to the autonomic (or vegetative) nervous system. This includes two sets of nerves, the sympathetic and the parasympathetic.* The former stimulates the organs at the body center only, and the latter, if ascendant, slows down the heart and breathing, but expands blood flow to the periphery by dilating the blood vessels. Health requires the full and free communication of impulses between the nerve filaments of the autonomic system, the brain and the voluntary nervous system. Huard points out, "The original aim of acupuncture was to suppress (energy) stasis and restore the normal *mobility* of the intravascular fluids and *tch'i*" (Huard and Wong, p. 209). This corresponds

---

* The parasympathetic system tends to induce secretion, increase the tone and contractility of smooth muscle and cause the dilation of blood vessels.

exactly to Reich's emphasis upon the vegetative system and the importance of free energy flow within it.

Medically, acupuncture concentrates on examining and probing the twelve fundamental meridians for an excess or deficiency of energy (an excess can be as serious as a shortage). In a sense these meridians, longitudinal lines with a nerve network, link the organs with the external skin surface. "Each of these twelve meridians extends along one of the limbs, one end being at the fingers or toes and the other at the face or chest: there are six upper limb . . . and six lower limb meridians" (Moss, pp. 30–31).

Moss tells precisely how the acupuncture doctor works.

The needle is taken between the thumb and the right-hand index finger. The skin over the acupuncture point is slightly stretched by using the index finger and the thumb of the left hand and the needle is inserted by clockwise rotation. The main object is to stimulate the affected organ through its meridian and restore the harmonious flow of energy which can be felt at the affected point, often within a minute. Once this is fairly normal, which should occur within a few minutes, the needle or needles (in other areas) can be withdrawn. When a needle has been inserted into deep tissues, where there is much muscle spasm—for instance in the buttock muscle—I usually keep it in position for a few minutes until I feel the tense muscle relax. The needle is then slowly withdrawn. [*Ibid.*, p. 40]

An unpublished paper by Canadian chiropractor William Morris claims there is now histological evidence for the acupuncture system. A revolutionary discovery by Dr. Kim Bonghan, of North Korea, made public in 1962, claims that acupuncture points are groups of small oval cells, surrounded by capillaries in the skin but also deeper within the body, extending to the internal organs. "This structure, a solid object, can be demonstrated just as a nerve ending, lymph node or any other structure can be under a microscope, by dissection, by staining with different chemicals" (Morris, p. 7). By classical histological methods of microtomy and staining, Bonghan showed there are two forms of special corpuscles, the superficial and the deep. The deep ones lie around the internal organs, the blood vessels and the lymphatic vessels. "They are linked to each other by ducts which have received the name of Bonghan ducts" (ibid., p. 8). Through slides and photographs Bonghan demonstrated that the ducts contain not cells but a liquid, free-flowing and acellular. Its circulation has been determined by the aid of radioactive tracers, using isotope phosphorus 32. Apparently it is a slowly flowing liquid, activated by the heart. When radioactive phosphorus is injected into a superficial duct, its path can be traced; it more or less follows the path of the meridians. Korea has named this bodily system the "Kyangrak" system. It is apparently a system on its own, as structurally complete as those of the vascular, nervous and lymphatic systems. Moreover, according to Bonghan, this fluid contains a large quantity of deoxyribonucleic acid (DNA) and ribonucleic acid (RNA) (Morris, p. 9). In Dr. Dimond's article, referred to earlier, he claimed that a Chinese medical team had been unable to confirm this Korean finding (p. 1163).

The exact location of acupuncture points has long been a matter of some difficulty. Moss says that the site of some is known to efficient masseurs, who call them nodules. He says: "These are sited below important points of acupuncture and consist of accumulated oedematous tissue that is congested with waste products . . . The points of Judo at the base of the neck consist of a similar build-up over some important acupuncture points . . . a sharp blow on this area can cause immediate unconsciousness" (Moss, p. 45).

When illness has been developing for some time, Moss says, the points of the affected meridian tend to become engorged, and by gently feeling them one can tell that the surrounding tissue is tense. The engorgement is due, he feels, to a deficiency of nerve impulse, leading to a slow-down in the circulation of the blood. In short, these sensitive engorged spots are apparently linked to energy stasis, and to ill-health. It is precisely these kinds of spots that the Reichian therapist or bioenergetic therapist will press, to get an emotional reaction—usually rage or fear—whose release is basic to orgone therapy as described in Chapter Two. Thus, bioenergetic therapy may be simply another way of trying to activate the acupuncture points and hence effect vital energy flow and health.* Called by some doctors "trigger areas" these sensitive spots, when pressed, may often trigger off relief of pain in another area of the body altogether; an area linked by a meridian to the trigger spot . . . (*ibid.*, p. 84).

In recent years the application of medical technology to acupuncture has opened up some interesting new vistas. According to Huard and Wong, French, Japanese and Chinese scholars (in addition to the Russians) have constructed equipment which has transformed acupuncture techniques by plotting the precise location of the points and demonstrating their differential electrical resistance. Also, these investigators have shown (a) that the skin on the line of the acupuncture points is more permeable to an electric current than is the nearby cutaneous zone, and (b) that the current does not pass through this zone in the same way in a pathological case as in a healthy one. Recent tests also indicate that medical acupuncture "acts upon the composition of the blood, the speed of sedimentation, the rates of hemoglobin, fibrinogen, agglutinins . . . the rate of coagulation, glycemia . . . the reticuloendothelial system, the endocrine glands, the neurovegetative system and the central nervous system. Its effect in the last-mentioned case has been recorded by systematic encephalography" (Huard and Wong, p. 212). Moreover, new techniques have recently been devised (De la Fuye) including "a combination of three factors, pricking, intermittent electric current and medicinal ions (ionizing acupuncture) which have given better results than acupuncture alone, especially in arterial hypertension" (*ibid.*).

Two specific devices that promise to make an important impact on medicine as well as on theory are worth describing. First, Dr. Richard Croon has perfected and marketed two elaborate machines, one, called an "Electroneurosomagraph," to locate and test the energy flow from the meridians, and the other to pump a low-voltage charge into those meridians which need normalization. Croon has outlined the work of these machines in a book, *Elektroneural-Diagnostik und*

---

* Some of the Reichian pressure is directed at more conventional points, namely, a few of the deep reflexes in the back and the cremasteric reflex in the side of the thigh, et cetera.

*Therapie.* He refers to the acupuncture points as Reaction-Sensor-Terminals (RST). These points differ from the general skin area by their strongly increased electrical conductivity. Croon makes use of AC at a frequency of *9,000 Hz** in order to measure the capacity and resistance at the points. Two electrodes are used, one being held in the hand of the subject. Using elaborate electronic equipment, the value of the Ohmic Resistance and Capacity is read off a dial or recorded by means of a magnetic arrangement on a chart. The recorded measures are then compared with the normal expected range, and this indicates to the doctor at which points there is an excess or lack of vital energy. With the other machine, called a "Perductor," an electrode is placed over the acupuncture point, and electricity is shot in, at a magnitude of approximately 100 resistance amps—that is, lower than the usual currents used in electrotherapy. When the scale reaches the normal figure for that point, the current can be automatically shut off. "The sensation of the patient on an individual point does not exceed a minor tingling and is not uncomfortable or painful. The apparatus is constructed so that the adjustments of the current can be made by the operator or the patient" (Croon, p. 6). Dr. Croon's machines are used in a clinic at Bad Homburg, West Germany, elsewhere in Europe, and also in cities of Canada and the United States.

The other machine was designed by Victor Adamenko, a Russian physicist, and its aim is to measure changes in the bioplasmic energy in the body by tapping the acupuncture points through a complicated electronic apparatus. "Variations in energy can be graphed and intensity of reaction shown on a numerical scale" (Ostrander and Schroeder, p. 227). This machine, called a "C.C.A.P." (Conductivity of the Channels of the Acupuncture Points), puts electrodes on the acupuncture points and picks up energic changes caused by varying emotional states and changes in consciousness, including hypnosis. Working with the Soviet psychiatrist, Raikov, Adamenko reported in the 1968 *Journal of Neuropathology and Psychiatry* of the prestigious Sechenov Medical Institute, that "with the C.C.A.P. we can chart objectively the psychical activity of the mind in states of somnambulism

* Hz = cycles per second.

and various levels of hypnosis" (*ibid.*, p. 150). What this indicates is that researchers working independently in Korea, Germany, Russia and elsewhere are all confirming the significance of energy flows for the health of the organism and that small differences on the skin link up to pathological energy transmission. Here is support for Reich's insights on muscular armoring and vegetotherapy and the case for an energy basis to health and forms of consciousness.

For an understanding of the medical practices of pre-Westernized Hawaii and their energy theory we are indebted to Max F. Long and his *The Secret Science Behind Miracles* and *The Secret Science at Work*. Max Long went to Hawaii in 1917 and with infinite patience and ingenuity gradually uncovered *Huna*, the secret medicine of Hawaii handed down from ancient times by their magicians called *kahunas*. Among the skills of the kahunas was the ability to perform miraculous or spiritual healings and to control the winds, the weather and to foretell the future. Long personally witnessed these things. Using a combination of massage, baths, bodily manipulation, suggestion and the laying on of hands, called *lomi-lomi*, they also performed extraordinary cures.

Before examining in detail their employment of *mana* it is essential to go into some of the fundamental aspects of kahuna psychology, the key point being that man is a trinity. A basic concept of the kahunas was that

> the low self (the subconscious mind) makes the vital force
> from our food, that the Middle Self (the conscious mind)
> takes this low-voltage force and steps up its voltage for its use
> in "will" and that the High Self (the superconscious) can take
> the force and step it up to the highest voltage, whereupon it
> can perform healing miracles. The "low-voltage" vital force
> . . . can carry chemical substances with it as it flows from
> person to person. It may also carry . . . thought forms. This
> notion throws new light on the power of suggestion when
> allied to the laying on of hands. This vital energy also can be
> stored in wood and other porous substances. A large discharge
> of this low-voltage force, commanded by the "will" can exert
> a paralyzing effect, or a mesmeric effect resulting in unconsciousness, sleep and the rigid or cataleptic state. [Long,
> *Secret Science Behind Miracles*, p. 138]

This might help to explain how certain evangelical healers, when they lay on hands, cause some "patients" to fall at once into a kind of coma.

Long observes that "some people have a natural ability to lay their hands on another who is weak or ill and cause vital force to flow into them from their own bodies, thus strengthening the patient. This is the simplest form of treatment with shared vital force" (ibid., p. 230). He is convinced that almost any healthy person can help the sick by laying on hands and making a willed command that the force enter the patient. He claims that "vital force, or mana, . . . responds to the command and direction of the consciousness of sentient beings almost as if it were itself conscious" (ibid., p. 230). Interestingly, Long also claims that "vital force is like the widow's mite—it increases as it is given." And by an "effort of the will one can cause the low self (subconscious) to create an excess supply of vital force. Almost anyone can learn to do this in a dozen lessons of twenty minutes each. When we have more vital force in our bodies than some other person and lay our hands on that other person with the will to give them vital force through our hands, the flow commences" (ibid., pp. 239–40).

Long also maintains that "flowing vital force causes a tingling sensation. This tingle . . . is a great assistance," he claims, "in determining whether or not the low self has obeyed the order and has made contact" (ibid., p. 325).

Long uses mana theory to explain what happens at séances when people sit "in circuit"—i.e., holding hands or touching knees. With no knowledge of L. E. Eeman's theories, Long argues that, in a materialization séance, the deceased spirit, lacking a physical body and its natural charge of vital force, gets

vital force and physical matter . . . from the circle of living sitters. The physical matter is changed to the thin ectoplasmic form and then solidified in the mold of the spirit's low shadow body.* This can result in a "full materialization" of an actual

---

* This is the "body" attached to or enveloping the subconscious. "It is an ideal conductor of the low mana and can be used as the storage place for it" (Westlake, Further Wanderings in the Radiesthesic Field, p. 11).

living, breathing, warm and completely normal physical body
. . . Such bodies have repeatedly stood close medical inspection. However, they are not permanent . . . the ectoplasmic
material (soon) is returned to the living and the solid form
vanishes. [*Ibid.*, p. 206]

(The evidence in the literature of parapsychology for material-
izations is fairly extensive, including hard-nosed researchers
like Hereward Carrington, in his *Laboratory Investigations
into Psychic Phenomena* [New York: McKay, 1939]; Baron
von Schrenck-Notzing, in *Phenomena of Materialization*
[London, 1925], and Dr. Glen Hamilton, a Canadian medical
man, whose fascinating experiments are found in *Intention
and Survival* [New York: Macmillan, 1942].)

What is the source of the kahuna's *mana*? It is related to
food, the bloodstream and the air. By breathing more deeply
one can increase its supply. When the kahunas wished to accumulate a surcharge of *mana*, "they breathed deeply and
visualized *mana* rising like water rises in a fountain, higher and
higher, until it overflows. The body is pictured as the fountain
and the water as *mana*" (Long, *Secret Science at Work*, p. 79).
Individuals with a low normal charge of *mana* can usually
sense the addition of the *mana* surcharge. It adds to their
sense of well-being and of physical strength, sharpens the
mind and makes the senses more acute, particularly vision.
Detail stands out and the colors get stronger.* Long says:
"All the evidence shows that the *mana* is . . . the life force,
and that with it the life is strong, while without it, it fades. This
is not exclusively a . . . Huna discovery" (*ibid.*, p. 82).
While some may say that these statements reflect but common
sense or logic, such claims for more acute vision and sense of
well-being recall Orson Bean's typical reaction to orgone
therapy. One comes much more alive as one either takes in
more *mana* (orgone) or allows it to flow more freely through
the body. I personally have experienced this after a few sessions of bioenergetic therapy.

In short, the kahunas' theory and practices suggest they
were dealing with ways of maximizing orgone energy. Thus,

---

* This reminds one of A. H. Maslow's "peak experiences."

it seems that an energy theory like the orgone not only was known to ancient civilizations, including the Greeks and the Hindus, but also was circulated among the Chinese, Koreans, Japanese and as far east as Hawaii. Perhaps, it was still more widely known but became lost with the victory of patriarchal civilizations, as Reich maintains in *The Mass Psychology of Fascism*.

*Chapter Nine*

# Magnetic or Psychic Healers

Bagnall, for all his tough-minded and skeptical approach, suggests that persons with strong vital organisms—and auras—possibly can pass on a slight healing force to others of less dynamic constitutions. Reich himself claimed that the orgone flows from the hands and tried to photograph such emissions. He was not vitally interested in that kind of healing but, as with dowsing, he believed the orgone theory had real explanatory powers.

Over the centuries there has been uncontrovertible evidence that a few individuals, through passes or laying on of hands, have effected some remarkable changes in the health of "believers." Many critics put all such healings down to the power of faith or suggestion. An alternative hypothesis is that faith and some healthy radiation were at work conjointly. If these healers could be shown to have possessed extremely vital constitutions, both Bagnall's suggestion and common sense would indicate that a force might have flowed from healer to patient. Moreover, if the life style of these healers showed evidence of drawing in unusual amounts of vitality, by practices such as meditation, deep breathing and a general life style of exceptional creativity, one would have further reason for seeing their healings as having a natural physical basis.

We will explore this idea by looking closely at the lives,

practices and thoughts of a number of contemporary healers of wide acceptance. In addition, we will look for ideas or evidence that corroborate orgone-flow and orgone-therapy claims.

Before examining directly specific modern healers, it will be useful to review briefly the method of magnetic healing as instituted by Mesmer and copied by unnumbered magnetic healers thereafter.* The basic technique was to sit near the patient and use the hands to touch or stroke the afflicted part or to "make passes" with the hands over the entire body from head to foot. Verbal suggestions were usually eschewed. The quality and duration of treatment was determined by the nature, extent and development of the individual case. The "healer" usually followed his intuitions as to how long to work and what "passes" to make.

Writing in the neo-Reichian journal, *Creative Process*, Jerome Eden tells how he taught magnetic healing to ordinary people in Alaska—where the nearest doctor was 120 miles away—and the results it achieved for painful conditions of various kinds. Two of the cases he refers to illustrate both the method and the kinds of effect that can be produced:

> A woman in her middle fifties had been suffering excruciating pain in the left shoulder and arm for several weeks. The diagnosis was arthritis, and a regimen of drugs and exercise had been instituted at a hospital, with no relief. The woman could not raise her arm above waist level, and she required help in dressing and washing. She was seated in a chair, her palms resting upon her knees. In accordance with Mesmer's method, gentle, slow passes were made from the shoulders to the fingertips, with pauses of 15–30 seconds over the site of the most critical pain areas. (The skin was not touched.) After 15 minutes, the woman was asked how she felt. Haltingly, she raised her arm. Her face expressed astonishment as her arm continued straight up over her head. All pain was gone. The woman declared it "a miracle" and

---

* It is interesting that Still, the founder of osteopathy, and Palmer, the founder of chiropractic were originally magnetic healers, and that a recent breakaway sect from chiropractic called "ontology" has reverted to a kind of magnetic healing (laying hands over, not on, the body).

ran to telephone her friends. To date, some two years later, she has had no return of symptoms. . . .

A 12-year-old girl fell on a flight of stairs and injured her right elbow. Immediately swelling developed, along with a painful redness about the entire elbow. The girl's father was greatly worried, because there was a history of osteomyelitis in the family. Adverse weather conditions made it impossible to drive the girl over the snow-covered Alaskan highway that evening. I gave the father a copy of Mesmer's *Maxims on Animal Magnetism*, with instructions to try Mesmer's method in an attempt at alleviating the girl's pain. Arrangements were made to drive her to the nearest hospital, weather permitting, the following morning. The next day, the father told me that he had witnessed something which "he still couldn't believe, even though he had done it himself." After his daughter had fallen asleep the previous evening, he had seated himself by her bedside. Taking her hand in his own, he had, with his other hand, gently stroked her injured elbow, just barely touching her skin with his fingertips, and increasing the length of the passes from her shoulder to her fingers. After a few moments he was amazed to watch the reddened area slowly being "drawn down" from the elbow area to her hand, and then "out of her fingers." He noted that his palm grew hot with what he described as "a glowing heat." After 30 minutes, all traces of swelling and redness were gone. And when she awoke the next morning the girl reported no sign of the injury. [J. Eden, "The Tender Touch," *Creative Process*, Nov. 1961, p. 85]

The basis for this might be orgonotic. But it also could be explained by simple blood mechanics. Parts of the body that are healthy have good blood circulation. Anything that will increase blood flow to an injured part will help. Thus, gentle massage or passes may be erotic in nature, and a vasodilatory parasympathetic response might follow with blood sent to the periphery.

It is valuable to note that many Christian thinkers, including Dr. Leslie Weatherhead, draw the clear distinction between all magnetic healing and that carried out sacramentally or through spiritual forces by saintly persons. In making this distinction they have the support of the great majority of both Christian and occult writers. This majority view is that persons with a certain physical constitution, usually including lots of energy, can perform magnetic healings. Their spontaneously developed or learned "technique" will be much like that of Mesmer. Their successes are typically limited to less serious complaints unless they can evoke a powerful curative suggestion in the patient, in which cases they may achieve quite astounding cures.

Characteristically, laying on of hands of "magnetic" or psychic healers depletes their energies to a greater or lesser extent. Dr. Weatherhead quotes a Dr. Petetin, of Lyons, and Deleuze, a French naturalist, for this view of healing. Deleuze, in his *Critical History of Animal Magnetism* (1813), wrote: "It is an emanation from ourselves guided by our will . . . He who magnetizes for curative purposes is aiding with his own life the failing life of the sufferer" (Weatherhead, p. 114). In the process such healers may even take on some or all of the symptoms of the patient. To avoid this, most of the books on magnetic or pranic healing advise the healer to shake or flick the hands and wrists at the termination of a healing, thereby hopefully "shaking off" the unhealthy emanations from the patient that could cling to hands and arms. Some even advise washing the hands (Yogi Ramacharada, p. 70). This "contamination" of the healer reminds one of how healthy people in Eeman's healing circuits would initially take on the fevers or other symptoms of their sick neighbor in the circuit.

The true saintly healer apparently disposes of a different and superior energy. He can heal at a distance as well as with physical contact, and sometimes virtually instantaneously; no serious illness (e.g., cancer) is beyond treatment; the healer is not susceptible to catching anything of the patient's sickness and he can go on for hours without tiring. In fact, he usually ends up a healing session feeling more energetic and spiritually resilient than when he began. (A conventional test of divine healing is whether it invigorates the healer.) Such saintly healers were fairly common in the first three centuries of Christianity and included a number of medieval saints like Saint Philip Neri.* Modern healers of this type

---

* For an account of these, one can read Dr. Evelyn Frost's *Christian Healing.*

would include William Hickson, of Britain, and Agnes Sanford and Katherine Kuhlman, of the United States. While it is possible that such saintly healers have an excess of natural vitality and may make use of this, their distinguishing quality is that they are spiritually very devout, see themselves as mere channels for a divine healing compassion and must keep themselves in tune with that Source or they are unable to function (Weatherhead, p. 139).

Our major concern here is the magnetic healing force. When Reich spoke of the hands emitting healing orgone, it seems that he was talking of this natural-energy force and not the other. Moreover, pancultural and historical analysis indicates the magnetic-healing ability is surprisingly widespread, while divine healers are very rare.

Since the First World War, Great Britain has seen much interest develop in spiritual healing, along with enthusiasm for the psychic and radiesthesia. France is said to have tens of thousands of magnetic healers (Pauwels and Bergier, *Dawning of the Magicians*). Weatherhead describes the penetration of healing into the nation's main churches through the formation of almost ten different voluntary associations. These, ranging from the Anglican Guild of Health (1905) and Guild of St. Raphael (1915) to the Friends' Spiritual Healing Fellowship and the Churches Council of Healing, all encourage the laying on of hands and quiet, unspectacular healing services. As early as 1930 a Committee on Spiritual Healing of the staid Anglican Church urged that healing practices of unction and laying on of hands "be used only in close conjunction with prayer and spiritual preparation." In the early fifties a Churches Fellowship for Psychical Study emerged and encouraged both research on healing and the practice of laying on of hands. By 1970 it numbered thousands of clergy and over twenty Anglican bishops among its members. So widespread is magnetic healing in Britain that the government in recent years officially licensed a number of healing homes and institutions in which healers work in conjunction with medical doctors.

Thus, regular clergy are finding they have something of a magnetic healing gift, and most seem to use it with great and understandable caution. Dr. Leslie Weatherhead tells, in his classic *Psychology, Religion and Healing*, of a Methodist minister whom he knew personally. This clergyman discovered his power accidentally when trying to comfort some parishioners whose child was supposedly dying. He simply put his hands on the child's head, with no thought of healing.

> Nothing was farther from this minister's thoughts than the laying on of hands. To his amazement, as soon as he laid them on the child's forehead, his hands trembled and shook as if with a clonus. Medically and surgically, all hope of recovery had been given up. The child stirred, opened his eyes, and then, with a sigh, went to sleep. He slept naturally and peacefully for some hours, and then, with the speed at which little children so frequently recover, he sat up, wanted food and toys and appeared to have recovered. In a few days he was well. [Weatherhead, p. 145]

This trembling of the arms is a very frequent sign of a healing activity. Another characteristic, also noted in Weatherhead's minister friend, is that patients claim they feel a "current like electricity" pass through that part of the body on which the hands are laid, or they feel intense heat.

Several unusual British medical doctors have for some years carried out a kind of spiritual healing as a supplement to their regular medical practice. One of these, Dr. Christopher Woodard, a Harley Street specialist, traveled widely throughout the Commonwealth in the fifties, conducting healing services; he also wrote several well-received books, including *A Doctor Heals by Faith*. A man of dynamic vitality and strong religious convictions, he aimed at the divine healings of the saints. Dr. Woodard assumes the existence of "a physical magnetic force" in certain healers that flows through the hands and says, "healers who feel exhausted [after healings] are, I believe, frequently using only some physical magnetic force and not the Divine Inspiration of healing . . . they are merely practising a phenomenon that can be easily explained on a physico-magnetic basis" (Woodard, p. 94). Many would doubt it is so easily explained.

Another London doctor with a reputation for effecting unusual cures in postwar Britain is Dr. Michael Ash. Dr. Ash claimed not only to heal, but frequently to be able to locate and diagnose the site of a disturbance by holding his hands a few inches from the skin of the patient. Dr. Weatherhead quotes him (without name) as follows:

He alleges that a healing-force flows from his own fingertips and can follow the "nerve paths" up to the injured tissue, or can win a muscular twitching reaction [a "spasm," in Reich's terminology] or response from say, a twisted pelvis, impelling the patient himself to correct the distortion by the involuntary contortions set up by the healer's hands. After watching him at work, after reading some of the literature, and after finding that when he held his hands a few inches from my own bare shoulders they subsequently burned all day as if they had been poulticed, although he made no suggestion of heat, I have myself become convinced that a *prima-facie* case warranting careful investigation has been made. [Weatherhead, p. 221]

A Russian healer who produces a similar heat is described in Chapter Fifteen.

One of Michael Ash's books is *Health, Radiation and Healing.* In it, he writes of his experiments to demonstrate the existence of a charge in the air around the living body. He claims he can make a rotary electroscope move slowly when the hand is held near it. This device was first used clinically to demonstrate the energy exchange during childbirth. He claims that a trained operator can feel a change in sensation in the palm of the hand when it is passed through a critical distance from the body examined. This phenomenon is diagramed in the book in terms of an experiment with hand energy and a watch. Compared with a control, Ash claimed that the energy from his fingertips, when they were held near the watch, caused it to gain time—some twenty seconds. A watch placed near a painful back in a short while lost time, in this case over ten seconds (Ash, *Health, Radiation and Healing,* p. 13).

Ash, though a convinced Christian, claims his healing method "has a purely scientific basis. Its use is limited by our lack of knowledge" (*ibid.,* p. 22). He describes how one can begin to acquire the necessary knowledge:

The would-be healer should start practising on some obvious condition. He should never interfere with correct orthodox

treatment but use his gift to enhance whatever is already being done for the patient . . . The simplest technique is to hold
the hands on either side of the place to be treated about an
inch from the skin. Move the hands slightly until you feel a
change in sensation. Go back and forth over this area till you
are sure you can feel the change again and again. Try to focus
your hands on the point of maximum sensation. Do not say
anything to the patient that would tell him what to expect, but
notice any reaction he makes to your efforts. If he feels anything anywhere relevant, check to see if you can feel anything
there as well. If you do feel anything, continue to use your
hands over the area to be treated till the sensation grows to a
maximum and then drops off suddenly. This is the point at
which to cease treatment for the time being. It is as if you
have induced a charge that has sparked from your body to the
patient's. Your hands and face may feel charged up negatively
as a result of this sparking. . . . To put yourself right again
you have to earth* your hands and face in water. The best
guide [on how often to treat the patient] is to observe how
soon the effect of your treatment wears off, and to give another
treatment just before this occurs in future. The gaps between
treatments can gradually be lengthened. As far as I know, you
cannot do the patient any harm by overdoing it, but you may
exhaust yourself. . . . This type of treatment stimulates the
body to heal itself. The response in each case differs according
to its present needs; and in any one case differs from time to
time. . . . This type of treatment given to burns and abra-
sions, cuts and ulcers, should merely quicken the natural
process of healing. . . . Just as in childbirth the mother's
body responds in such a way as to make itself comfortable by
pushing out the child, so does the body respond under the
stimulus of the healer. . . . As a routine, it is best to treat
through the spine—starting from its base and working up towards the head as the sensation that you feel moves up the
spine. . . . Apart from the spine and the top of the head
where the body as a whole is represented, and where general
healing treatments can be given, the hands and feet are good
points of contact. [*Ibid.*, pp. 23–26]

---

* This means to discharge the "diseased" energy.

Several of these statements seem to jibe with or support elements of the orgone theory. For instance, the idea of inducing a charge into the patient's body, the notion that such treatment strengthens the body's natural healing powers, the idea of the body making convulsive moves to push out the alien or toxic matter, and the suggestion of working up the spine from the base, thus following the body's segments. In fact, one would imagine mistakenly that Ash derived some or many of his proposals from a study of orgone energy.

The most widely known healer of Britain, if not the entire British Commonwealth, is spiritualist Harry Edwards, who began his ministry in the thirties. Edwards, who has written numerous books on healing and was for years president of the National Federation of Spiritual Healers,* has made a considerable impression on the British medical scene. He has been interviewed on TV, held packed healing meetings at Albert Hall, London, and developed among spiritualists a large company of fellow healers. (All together, in Britain there were in the sixties over a thousand spiritualist healing groups.) In 1960, a struggle occurred with the British Medical Association, who tried to have the Federation's healers banned from attending patients in hospitals. However, the Minister of Health decreed that when a patient asked for a healer and the hospital staff had been consulted, the healer could function in that hospital. In 1960, some 264 hospital authorities gave permission for healers to act, upon request; this involved nearly two thousand hospitals.

In reference to the source of healing power, Harry Edwards sounds strikingly like Wilhelm Reich, although he arrived at his theories quite independently:

We live in a "sea" of cosmic energies. We unconsciously absorb these into our bodies according to our needs. . . . Apart from the food we eat we need cosmic energy for our well-being. While this is naturally and unconsciously absorbed, we can directively take it into ourselves through intention, but it needs confidence and belief in the act to do so. . . . Cosmic

* Among its patrons in 1961 were Sir John Anderson, Bart; Sir Adrian Boult; Lord and Lady Dowding; Lady Fforde; J. Arthur Findley, MBE; Brigadier Fireplace; Lieutenant Colonel Reginald Lester; Beverley Nichols; Mrs. Gladys Osborne Leonard; Lady Maud Duckham.

energy belongs to that part of our realm that is near-physical;
that is, the inter-state between physical and spirit. [Harry
Edwards, "Cosmic Strength from Trees and Water," *The
Spiritual Healer,* May 1956, pp. 197–98]

In an article on the prevention and cure of cancer, Edwards,
after citing considerable medical research, writes: "Thus I
would like to sum up, by saying that there is an accumulating
mass of evidence that cancer has a psychosomatic cause. The
removal of these causes is generally outside the scope of the
physician. They do lie within the influence of spiritual healing" (Edwards, in *The Spiritual Healer,* September 1955, pp.
420–21).

The belief that most cancers have a psychsomatic origin
is very common among spiritual healers. Usually they attribute the emergence of cancer to persisting resentment, hate
or allied emotions, which have disrupted interpersonal relationships and poisoned emotional satisfactions. The result is
that people are out of harmony with both themselves and
those close to them. Thus, it is natural for Harry Edwards
to quote from an article by Sidney Katz, "Does Worry Cause
Cancer?" in Canada's *MacLean's Magazine,* March 5, 1955:

> "We are confronted with the possibility," says Dr. Phillip
> West, a University of California professor of biophysics, "that
> all of us may have had or will have some form of cancer but
> because of inherent natural control of this process, we will
> never know it." . . . Research . . . shows that those patients
> who had developed cancer in an active way were made up of
> individuals who were painfully sensitive, over-nice, apologetic
> and over-anxious to please; also the frustrated, vicious and
> those who had carried revengeful thoughts, hatreds and selfish
> desires. In all these conditions there is a bottling-up of feelings
> with no way of releasing the mental tensions. [*The Spiritual
> Healer,* September 1955, pp. 419–20]

The similarity of this analysis to Reich's and its emphasis
upon emotional blocking or stresses are very interesting.

One American counterpart of Harry Edwards, in dedication
and extensiveness of healing impact, was Ambrose Worrall,*

---

* Mrs. Agnes Sanford, wife of an Episcopal priest, is another widely
respected spiritual healer.

ably assisted by his clairvoyant wife, Olga. They were doing healing for forty years and were well known in much of the United States through their lectures, radio and TV appearances. In their last book, written with Will Oursler, Ambrose Worrall differentiates the heat that accompanies the laying on of hands and the usual warmth of the hand: "People feel it as though you had put an electric pad upon their bodies" (Worrall, *Explore Your Psychic World*, pp. 97–98).

But the intensity of the heat felt is not necessarily linked to the chance of a healing occurring. In discussing their sensations during healing the Worralls said:

> To me it feels like warm air coming out of my hands. However, a healer can only explain what he feels. What he feels may be just the side effects. . . . Sometimes a person will say, "I just feel something penetrating like an electric current, and it's hot, it's burning me." . . . If a person is not sensitive to the energy being transmitted he may not feel it, but I can feel it coming out of my hands. . . . I feel the electric-like current (Olga) . . . like pins and needles. You have all touched a live wire at some time or other. You know the shock you get from it. That's what I feel in my hands. (Ambrose): It's not as powerful as the sensation experienced when bridging 117v 60-cycle electrical supply lines! (Olga) Oh no. But the prickly sensation is what is felt. (Ambrose): It's as if you took the terminals of a battery and touched them to your tongue. . . . You get a little tingling sensation, and that's what you often feel in your hands when you put them on a patient—but not always. It seems to vary in potential, but it makes me believe there is an electrical component involved. [*Ibid.*, pp. 98–99]

Julius Weinberger, an electronics expert, associated with Wainwright House, Rye, New York, tried a test on Ambrose Worrall. The latter describes it as follows:

> Julius Weinberger, an electronic scientist, devised the method used. A piece of dental X-ray film was placed on the palm of my hand with a lead bar across it; the whole apparatus was attached with adhesive tape. I asked, "What is that for?" Dr. Weinberger said, "You say you feel this force flowing

down your arm and out of your hand, and we think it might build up against the lead bar and maybe it will show something on this film." There was an ill woman in the next room who had been brought in to participate in the experiment. Dr. Weinberger asked me to put my hand on her and tell him if I felt the power flowing. This I did. They took the X-ray film from my hand, developed it, and there appeared a line of light on the film where the bar had been. They made some checks which indicated that the line was not caused by X-ray. I don't know whether they pursued it enough to find out what energy was involved, but . . . it takes only a few electron volts to affect X-ray film. [*Ibid.*, pp. 99–100]

To make this experiment scientifically impressive they should have tried the same procedure in a situation where Worrall did not feel the tingling . . . then look at the film. That would be a control. Or take a person who is not a sensitive and compare the results.

Another experiment was carried out in which a Sister Justa Smith, biochemistry professor, at Rosary Hill College, Buffalo, compared the effect on enzymes in distilled water held in Worrall's hands to the effect on them of magnetic fields.

Sister Justa said she got something called a significant change when I treated enzymes by holding my hands in their vicinity. She said the change far exceeded any resulting from the effect of the physical magnetic field . . . Sister Justa had spent three years studying the effect of electromagnetic fields on the rate of hydrolysis of a synthetic substrate catalyzed by the enzyme trypsin. She had shown that these magnetic fields produced a statistically significant increase in the yield of this enzyme reaction when compared to a control with no magnetic field. Then she had Ambrose Worrall hold a glass jar containing the enzyme trypsin and two other components. The result was an increase in yield of about twelve per cent . . . It had to be over ten per cent to be significant. [Elton S. Cook and Sister Justa Smith, in Barnothy, pp. 126–27]

In a subsequent study by Sister Justa, with a Montreal healer named Colonel Estebany, it was established that a comparable effect was produced by this healer's hands. Every day during the Estebany experiments, new solutions of trypsin were prepared and were divided into four stoppered aliquoits. One was retained in the native state and used as a control; the second was "treated" by the Colonel, who placed his hands around the bottle for a maximum of 75 minutes (during which time, portions were pipetted out after 15, 30, 45 and 60 minutes in order to measure any change of activity during these time periods); the third was "damaged" by ultraviolet light sufficient to reduce activity to 68–80 percent, and then treated by the Colonel as above; the fourth was exposed to a high magnetic field of 8,000–13,000 gauss for hourly increments up to three hours.

Three to five activity measurements were made on each of the above samples for each time interval. The mean and the standard deviation from the mean for each were calculated daily and, finally, a mean of the means and standard deviation over all the days were calculated for each sample. These were compared.

The results of these experiments indicated a significant increase in the activity of the "treated" enzymes compared with controls. (In case heat were a factor, the control was maintained at the temperature of Colonel Estebany's hands.) The activity of the enzyme in bottles held by the Colonel increased to a degree comparable to that obtained in a magnetic field of 13,000 gauss.

Sister Justa says, "It is interesting to note that the qualitative effect of high magnetic fields and of the hands of this healer are the same and they are also quantitatively similar up to one hour of exposure" (Rindge, "Are There Healing Hands?" p. 19).

In other words, healing energy has both light- and growth-promoting properties. Reich claimed the same thing for the orgone—that it could be photographed and that it could induce growth in plants.

Obviously, the Worralls, with their psychic and healing powers, are rather different from the average magnetic healer. However, it is possible that they simply give forth an exceptional charge of the same kind of energy. In an interesting article in parapsychologist Allan Angoff's book, *The Psychic Force*, Eric Cuddon, a British barrister with magnetic-healing abilities examines his capacity to relieve pain—usually in several minutes—by laying fingers on the affected part. Above all he stresses the role of self-confidence in effecting any appreciable results. He examines carefully the hypothesis that suggestion alone explains the effect. Then he adds:

> The hypothesis that makes most sense to me is that of electrostatic charge. I first conceived this idea when I treated a patient who was hospitalized. In this case, the man's whole background appeared to support the hypothesis of the restoration of electrostatic balance. He was a farmer in County Galway and it was usually windy on his farm. He began to get tenseness in his head and forehead only after being in the open air for a couple of hours. He suffered from a dry skin. Now, the two basic factors that favor the induction of a static charge are dryness and unidirectional friction. Might it not be possible, I reasoned, that a wind blowing across a dry forehead could produce a static charge on the surface of the skin, thus creating a tension of the nerve endings on the underside of the skin and resulting in a headache? [Angoff, p. 104]

Mr. Cuddon then devised an experiment to test this hypothesis:

> I used brown paper, dried by warming, and stroked it unidirectionally with vigor by means of a clothes brush. The charge produced thereby was sufficient to cause the paper to adhere strongly to my flat palm, and when the ends of the paper were bent away from the palm the paper would flip back instantly upon release as though activated by springs.
>
> The reason for the strong adherence of the paper is, of course, the tension set up between the charged surface of the paper and the palm of my hand, as a result of the dissimilarity of the electrostatic charges of the two surfaces. . . . If, having charged the paper, one places the fingers of the remaining hand (the paper adhering to the palm of the other) over the entire surface a little at a time, it takes just about two minutes for the charge to leak away, whereupon the paper will drop from the palm of the other hand. Similarly, a roughly equal amount of time is required for the charge to leak away if, instead of the fingers of the free hand, a piece of

metal previously grounded is passed to and fro across the charged surface of the paper. The time this procedure takes is roughly the same as the length of time needed to produce relief from pain by laying on of hands. [*Ibid.*, pp. 104–5]

Mr. Cuddon theorizes that some of his healings were due to an equalizing of static charges. He defines static charge:

> It is a difference of potential as between two points caused by the one having a paucity or superfluity of electrons relative to the other. Thus, quite apart from any friction on the surface of a dry skin, it seems possible to me that certain changes in the electrical potential of the nerve ending *under* the skin might be caused by worry in the mind of the patient, or for other reasons. If such resultant potentials were greatly different from the potential at the *surface* of the skin, this circumstance would cause tension and possibly pain, say, in the forehead, or at any rate might be a contributory factor. [*Ibid.*, p. 105]

This explanation requires postulating something causing a dissimilar potential under the surface of the skin. This might be tense musculature, or other aspects of muscular armoring. Regardless, this theory is really interesting, as it admits of some simple physical test and is rooted in a concept of energy flow.

Evidence seems to be accumulating that magnetic healing is a natural, physical phenomenon involving a transfer of energies subject to scientific measurement. Those regularly engaged in effective healing generally possess a surplus of natural vitality and radiate a quiet self-confidence. They may be likened to reservoirs of unusual bio-energy. In Chapters Twelve and Fifteen, quite recent studies of healers underline these observations. In the latter chapter, where a Russian healer, Colonel Krivorotov, is discussed, evidence of the operation of an electrostatic field during his healing is put forward. Mr. Cuddon's ideas may therefore be taken seriously. It could be that further testing of this component of magnetic healing will illuminate the whole subject while throwing fresh light on the orgone theory.

PART

IV

*Chapter Ten*

APPLICATIONS OF ORGONE TO HUMAN
ILLNESS AND AGRICULTURAL EXPERIMENTS

One of the most distinctive differences between Reich and
earlier discoverers of life energies lies in the much greater and
more diversified application he made of the energy. Mesmer
and his followers confined themselves to limited therapeutic
uses, Reichenbach was content to try to demonstrate the
existence of the od, and most subsequent investigators have
kept to limited healing or growth-enhancing applications.
Only the yogis have considered extensive nontherapeutic uses
of their prana, but here its utilization has generally been confined to a very few advanced adepts. Reich, on the other hand,
came to visualize more and more extensive and regenerative
functions for orgone. This grew naturally out of his early
concern for humanist, radical and political reforms. Somehow,
cosmic orgone energy provided not only possibilities for the
alleviation of cancer, the control of weather and the reclama-
tion of deserts, but even some nullification of the worst effects
of atomic radiational damage. While the validity of the orgone
as a real energy does not hinge on its successful application in
every one of these disparate fields, the accurate placement of
Reich in any hierarchy of modern discoveries does follow upon
the verification or repudiation of such claims.

Indeed, it is the very extravagance or breadth of his claims
that lent some weight to the oft-repeated charge that in his

later days Reich became quite paranoid. Since the judgment of the court psychiatrist at his trial and of a number of close associates in the early fifties—men well trained in psychiatric diagnoses—rejected this view, one must suspect those who bandied it about. In fact, this habit of labeling an extreme innovator as crazy or paranoid has been the rule rather than the exception in the history of great new discoveries. It can be seen as a natural sociological process concentrated among individuals and groups closely tied to institutions threatened by radical new ideas. In brief, the threat to their vested interests leads them to denigrate vulnerable innovators.

It is appropriate now to describe Reich's various applications of orgone, to evaluate their success as far as that is possible at the present time, and to see what these experiments and applications reveal as to the real nature of this energy. Our over-riding concern is not to pass any final judgment on individual applications, but to explore their implications for the validity of Reich's conceptualization of the orgone as a primordial life energy.

First, it is valuable to examine in some detail how Reich conceptualized cancer—and other degenerative diseases—and, how he and his followers experimented with orgone accumulators both to treat cancer and to prevent its outbreak. The story is told principally in his *The Cancer Biopathy* and in articles in the *Orgone Energy Bulletin,* one of the chief journals he published to describe orgone researches. It was precisely this work with cancer, and the rental of accumulators, that brought Reich into conflict with the U.S. Food and Drug Administration and led to his eventual imprisonment and two separate burnings of his books.* Thus, up until 1971, American publishers have been loath to reprint books and booklets directly concerned with the cancer researches, so that a booklet, *The Orgone Accumulator,* and his book *The Cancer Biopathy* are still in very short supply.

This book is one of the most important of Reich's contributions, because of its biophysical and social concepts and arguments. I believe that future research will vindicate a good many of its observations and general conclusions. This is *not*

_____

* Others were to be withheld from circulation.

to say that everything in it is empirically or theoretically sound. Reich repeatedly declared that many of his observations and concepts were tentative, unfinished, and subject to modifications. Even the most cursory reading of the book reveals in Reich the dedicated scientist's willingness to be led by the facts, to hold opinions tentatively and to faithfully report negative as well as positive experimental results. Preliminary hypotheses, such as the mechanism of the orgone's therapeutic affect, were later rejected for "a better explanation." Reich constantly uncovered new facts and seems ever willing to modify early conclusions. In addition, we note a medical man's natural caution in widening the use of new therapeutic techniques, the refusal to promise sure results (for accumulator treatments), a keen sense of the possible role of suggestion in therapy and a recurring use of controls in basic experiments. While Reich did not always write up his experiments in the conventional scientific style, occasionally omitting useful details and neglecting scholarly documentation or extensive replication, he did exemplify the essential scientific interest in making his theory correspond with the facts.

To understand his use of orgone in cases of cancer, one must first appreciate the meaning and import of his concept, the biopathy. It was in the treatment of biopathic or degenerative diseases that he saw the major medical contribution of the orgone. Rejecting terms such as "cancer disposition" as functionally inadequate, Reich insisted that the degenerative diseases are biopathies—that is, pathological developments in the biological process. "Under the term *biopathies* we subsume all those disease processes which take place in the autonomic apparatus. There is a typical basic disturbance of the plasmatic system which—once it has started—may express itself in a variety of symptomatic disease pictures" (*International Journal of Sex-Economy and Orgone Research*, Vol. I, 1942, p. 131). It may result in a cancer

(cancer biopathy) but equally well in an angina pectoris, an asthma, a cardiovascular hypertension, an epilepsy, a catatonic or paranoid schizophrenia, an anxiety neurosis, a multiple sclerosis, a chorea, chronic alcoholism, etc. What determines the development of a biopathy into this or that syndrome, we do not as yet know. What . . . all of these diseases have *in common: a disturbance of the biological function of pulsation in the total organism.*

A fracture, an abscess . . . yellow fever, rheumatic pericarditis, acute alcohol intoxication, infectious peritonitis, syphilis, etc., are *not* biopathies. They are not due to a disturbance of the autonomic pulsation of the total vital apparatus; they are circumscribed, and, if they result in a disturbance of the biological pulsation, they do so only secondarily. We speak of biopathies only where the disease process begins with a disturbance of the biological pulsation, no matter what secondary disease picture it results in. Thus, we can distinguish a "schizophrenic biopathy" from a "cardiovascular biopathy," etc. [*Ibid.*, pp. 131–32]

Cancer, Reich points out,

lends itself particularly well to a study of the basic mechanisms of biopathy. In it, we find a great number of disturbances . . . It shows pathological cell growth; one of its essential manifestations is bacterial intoxication and putrefaction; it is based on chemical as well as bio-electric disturbances in the organism; it has to do with emotional and sexual disturbances; it results in a number of secondary processes—such as anemia—which otherwise form disease entities by themselves; it is a disease in which civilized living plays a decisive role; it is of concern to the dietician as well as to the endocrinologist or virus researcher. [*Ibid.*, p. 132]

The basic stance taken by Reich was that "living functioning in man is basically no different from that in the amoeba. Its basic criterion is biological pulsation, i.e., alternating complete contraction and expansion" (*ibid.*, p. 132). This is observable in unicellular organisms in the form of rhythmical contractions of the vacuoles, in the contractions and serpentine movements of the plasma or in the pulse beat of metazoa. In various organs it takes different forms; for example, in the intestines we see it in the alternating contraction and expansion known as peristalsis. Based on microscopic observations of transparent worms, Reich refutes the common notion that the autonomic nervous system is immobile. (He contends that the ganglia fibers actually move.) In short, nerves and all parts of the organism move or pulsate until death, when a final rigid contraction—rigor mortis—sets in.

The operative source of a biopathy is found in biopathic shrinking. This begins

> with a chronic preponderance of contraction and inhibition of expansion in the autonomic system. This is most clearly manifested in the respiratory disturbance of neurotics and psychotics: the pulsation (alternating expansion and contraction) of lungs and thorax is restricted; the inspiratory attitude predominates. Understandably enough, the general contraction ("sympatheticotonia") does not remain restricted to individual organs. [*Ibid.*, p. 133]

> When it is serious, this shrinking extends to whole organ systems, their tissues, the blood system, the endocrine system as well as the character structure. Depending on the region, it expresses itself in different ways: in the cardiovascular system as high blood pressure and tachycardia, in the blood system as shrinking of the erythrocytes (formation of T-bodies, poikilocytosis anemia), in the emotional realm as rigidity and character armoring, in the intestines as constipation, in the skin as pallor, in the sexual function as orgastic impotence, etc. [*Ibid.*, p. 133]

If the chronic contraction of the autonomic system does not change or subside, a biopathic shrinking occurs, a characteristic result of which is cancer. With his typical focus on sexual expression, Reich considers the stasis, or locking-in, of sexual energies a corollary of the biopathic shrinking. The link between cancer and sexual function is seen to rest upon (1) poor breathing, including weak respiration; (2) disturbances in the discharge of sexual organs leading to disturbed orgonotic charges; (3) severe muscular tension (chronic spasms); and (4) chronic inability to let go sexually. "Muscular hypertension due to sexual stasis regularly leads to a diminution of orgonotic sensations; the extreme degree of this is the sensation of the organ 'being dead.' This corresponds to a block of biological activity in the respective organ" (*ibid.*,

p. 134). For example, the blocking of sexual excitation in the genitals is accompanied by chronic tight pelvic musculature and this is illustrated by the uterine spasms common in frigid women. Muscular spasms due to inhibitions of bio-energetic currents also occur, for example, at the throat, at the entrance to and exit from the stomach and at the anus, etc. It is at these places that cancer is frequently found. Chronic tension around glands, the skin surface or a mucous membrane area can prevent biological energy from charging these vital areas and can lead to trouble. The widespread sexual inhibition of women explains, says Reich, the prevalence of cancer in the breasts and genital organs.

> Many women who suffer from genital tension and vaginal anaesthesia complain of a feeling that "something is not as it should be down there." They relate that during puberty they experienced the well-known signs of biosexual excitation; and that later they learned to fight these sensations by way of holding their breath. Later, so they relate in a typical manner, they began to experience in the genitals a sensation of "deadness" or "numbness," which in turn frightened them. As the vegetative sensations in the organs are an immediate expression of the actual biological state of the organs, such statements are of extreme importance for an evaluation of somatic processes. [*Ibid.*, pp. 134–35]

A basic key to the biopathic disease is the breathing process. The individual who gets cancer, Reich claims, fails to effect full expiration; only part of the air intake is released. Proper breathing involves an involuntary sinuous movement of the torso and some sexual excitation in the genital region at the end of the expiration. Most people inhibit this by a pulled up chest and a tense abdomen. They hold in. The overall effects of such breathing, he adds, "can as yet not even be guessed at. Deficient external respiration must of necessity lead to a deficient internal respiration of the organs, that is, a deficient supply of oxygen and [deficient] elimination of carbon dioxide" (*ibid.*, p. 135).

To underline this critical observation, Reich quotes Otto Warburg, who found all cancer-producing stimuli tend to produce local oxygen deficiency, so that there is a disturbance of respiration in the cells. In short, the cancer cell is a poorly breathing cell. Warburg considered it to be a cell which developed out of adjustment to this respiration disturbance and the consequent weakened energy metabolism.

The specific contribution of orgone-energy irradiation, through the accumulator, is to help restore fuller breathing, and to overcome the chronic contraction syndrome of the typical biopathic disease. Expressed medically, "The orgone energy has a vagotonic effect, that is, it counteracts the general sympatheticotonic contraction of the organism" (*ibid.*, p. 139).

The following case of a woman with cancer of the left breast, who had been given at most two months to live and was treated in the accumulator, illustrates much of the theory. The muscular armoring was visible in an immobile chin; the woman talked through her teeth as though hissing. Her jaw muscles were rigid, as well as both the superficial and deep musculature of the neck. She held her head somewhat pulled in but thrust forward, as if afraid that something would happen to her neck. Her fifth cervical was collapsed and she had been wearing a plaster collar for some time.

Biophysically, while her reflexes were normal her respiration was severely disturbed. The lips were drawn in and her nostrils a bit distended. The thorax was immobile and failed to participate in respiration. When asked to, she was unable to breathe out deeply; in fact, she couldn't understand the question. Her head, neck and shoulders formed a rigid unit; she could move her arms only slowly and with effort. Her handclasp both left and right was very weak. The scapular muscles were extremely tense, the muscles between the shoulder blades sensitive to the touch. The abdominal wall was tense, reacting to the slightest pressure with resistance. The pelvis was immobilized and retracted.

The woman's history revealed that she had been married two years when her husband died. Early in the marriage she had been much excited sexually, but the husband was impotent and left her quite unsatisfied. Her lack of sexual gratification in the first few months was followed by a period of "getting used to it." Now she had been a widow for twelve years, devoted herself entirely to her child, had no contact with men and gradually withdrew from social contacts. Her sexual excitation gradually subsided and she developed anxiety states. "She appeared emotionally balanced and somehow reconciled to her sexual abstinence and personal fate. . . . She presented the picture of neurotic *resignation* with which the character analyst is so familiar; she no longer had any impulse to change her life situation" (*ibid.*, p. 138).

She was given orgone treatment daily in the accumulator for thirty minutes. During the first session the skin between the shoulder blades became red. Pain decreased in the back region. In the second session, the redness spread to the upper part of the back and chest, but disappeared as soon as she went out of the accumulator. With the third session she felt that the air was "closer and heavier" and said, "I feel as if I were filling up" and "something clears up in my body." In this session she began to perspire, particularly under the arms, though she had not perspired for years.

In commenting on this case, Reich points out, "all these reactions to the orgone radiation are typical of cancer patients. In one patient, one reaction will predominate, in another a different one. Such phenomena as redness of the skin, lowering of the pulse rate, warm perspiration and the subjective sensations of 'something in the body getting loose, filling up, swelling, etc.' admit of only one interpretation: the cancer habitus is determined by a general sympatheticotonia" (*ibid.*, p. 139). Most cancer patients, says Reich, show rapid pulse, pallor, dryness of the skin, reduced motility of the organs, constipation and inhibition of the sweat glands. In the accumulator, the pulse tends to come down—e.g., from 120 to 90—without any medication. Peripheral blood vessels dilate, the blood pressure decreases, and there is redness of the skin and perspiration. Thus, he contends, "Expressed in terms of biological pulsation, this means that the plasma system relinquishes the chronic attitude of contraction and begins to expand vagotonically" (*ibid.*, p. 139).

The typical cancer pain also decreases. Reich calls it a shrinking pain. He sees much cancer pain as due *not* to a local mechanical lesion but to a general "pulling" at the tissues, associated with the general contraction of the organism. With anxiety, he claims, we feel a crawling back into the self, a shrinking and a tightness, whereas with pleasure, there is a stretching-out toward the world, including an expansion or stretching-out of the autonomic nerves. Thus, when sitting in the accumulator, and taking in the orgone, one experiences a stretching-out, or expansion, of the plasmatic system, and the pain associated with "shrinking" disappears. In sum, he claims, "an essential effect of the orgone energy; it charges living tissues and causes an expansion of the autonomic nervous system" (vagotonia) (*ibid.*, p. 140).

After the initial improvement of the female cancer patient, the subsequent sequel of difficulties laid bare the complexity and depth of her cancer biopathy and its resistance to cure. After some ten weeks of orgone therapy the woman had apparently regained physical health; the tumor had gone, the blood tested healthy, she was up and working, to the sheer amazement of friends and family. Then, when she failed to "locate" a boy friend, while developing long-lost sexual excitations, she began to fail, both physically and somatically. "A general vegetative disease picture appeared which had been hidden and which formed the actual background of the . . . shrinking biopathy" (*ibid.*, p. 142). Over the next six months, phobias of sexual intercourse "breaking her spine," weakness in the legs, and a recurring anxiety about falling, pressure and then hypochondrial pains in the chest, spastic contractions of the diaphragm, bodily immobilization (hardly able to move her arms), a fear of dying, and finally the reappearance of new tumors occurred, in this sequence. While Reich's visits to her and reactivation of deep respiration helped temporarily and renewed orgone treatments produced a reappearance of some health and mobility, an accident to her left femur put her in the hospital, where she died in four weeks. Reich's conclusion was that orgasm anxiety and her inability to shed basic characterological problems and enjoy sex accentuated her sexual stasis and led to the return of cancer and death. This apparently typical case demonstrates that where a shrinking biopathy is well established, and the disease process medically irreversible, orgone irradiations— even if combined with some vegetotherapy—are unlikely to lead to a cure.

Other cases reported in *Cancer Biopathy* indicate another

basic problem. Where patients have, for example, a medium-sized growth on the breast which disappears after daily orgone irradiation, a considerable number are unable to eliminate the toxic matter from the dissolving cancer. In these cases, they may die of kidney or liver breakdown (or they may live months beyond medical prognosis and feel no pain, but their biological systems, being unable to eliminate the toxins, collapse at one vital point or another). This kind of result was detailed in numerous cases in *The Cancer Biopathy*, and yet Reich was charged by the U.S. Food and Drug Administration with mislabeling the accumulators as a cancer cure! The fate of the patient in orgone treatments, Reich explicitly states, "depends largely on whether these enormous masses of disintegration products [the tumors] can be eliminated from the organism or not" (*ibid.*, p. 276).

The precise way in which orgone irradiation breaks down the tumors—or otherwise acts therapeutically in biopathies—merits some description. Reich's explanation begins with the claim that cancer cells arise from tissue degeneration. Putrid degeneration of nonliving or living protein produces, he asserted, small black bodies of elongated shape, pointed at one end, which he called T-bacilli. They are weak in orgone charge and caused disease when injected into healthy organisms—for example, mice. (At a further stage they change into the clublike, caudate-form cells of fully developed cancer.) SA bions (from sand) which possess a strong orgone (blue) charge can either immobilize or kill the T-bacilli. Human blood can contain either T-bacilli or large blue bions and through a simple process, one can identify their concentration.* Orgone irradiation acts on the blood, conveying an atmospheric orgone charge to the red blood cells through the lungs. As these red blood cells circulate, they attract and then immobilize or kill T-bacilli, but in the process these cells gradually lose their orgone charge. They also act to dissolve cancer tissue into T-bacilli (*ibid.*, p. 247). In short, by introducing orgone into the body through an accumulator, the organism is relieved of the burden of having to use up its own orgone (in

---

* The process involves autoclaving a few drops of blood in bouillon and Kcl solution (potassium chloride) for a half hour at 120 degrees Centigrade and 15 pounds pressure.

red blood cells) in a fight against this disease. Those interested in more elaborate biological interpretations of the cancer cell and the therapeutic role of the orgone should consult *The Cancer Biopathy*, Parts VI and VII.

Having examined, if only briefly, the core of Reich's interpretation of orgone irradiation, we should consider the various applications of the orgone to disease states and the early results Reich achieved, as well as those of his disciples and others.

Shortly after discovering the atmospheric orgone, in July, 1940, Reich switched from treating cancerous mice with SA bions to putting them in his first accumulator. They were given a half-hour's irradiation per day, with the following results:

> The very first tests revealed an astoundingly rapid effect: the mice recuperated rapidly, the fur became smooth and shiny, the eyes lost their dullness, the whole organism became vigorous instead of contracted and bent, and the tumors ceased to grow or they even receded. At first, it seemed astounding that a simple cabinet, consisting of nothing but organic material outside and metal inside, should have such a pronounced biological effect. Later on, at a time when this effect had long since become a matter of fact to us, we saw this astonishment in many visitors to our laboratory. They looked for electric wires and complicated mechanical contraptions and could not understand that such a simple arrangement should be capable of influencing cancer. [*Ibid.*, p. 259]

In these mice the life span was increased over the controls by two and one half months (from an average of seven to nine and one half months after injection of cancer tumor).

Early applications of the accumulator on humans in 1941 revealed some basic principles. First, its inner walls must not be more than four inches distant from the organism or else the effects will be minimal. To explain this, Reich argued that the accumulator set up a force field that excited and interacted with that of the organism (mice or human) in the accumulator. If they were too far apart the two energy systems made little or no contact. Secondly, he found that lively individuals felt the orgone effect much more quickly and intensely than the vegetatively sluggish. He reasoned that the former had a wider energy field (aura). Thirdly, the latter usually failed to feel any effect until after a number of daily sessions. The explanation here was that the accumulator slowly charged these organisms and gradually extended or vitalized their energy field. In addition, the body's temperature increase in the accumulator varied with the individual. The typical slight rise in body temperature he ascribed to an increase in metabolism. Besides, "many patients who suffer from a standstill of their biological energy metabolism develop sexual excitation . . . under the influence of the orgone treatment" (ibid., p. 269).

The first cancer patients accepted for treatment by Reich's Orgone and Cancer Research Laboratory had to provide a signed statement that the physician in charge considered their case hopeless, and that neither the doctor nor the relatives were opposed to this experimental treatment. No fee was charged, and no promise of recovery was given. As indicated earlier, Reich admitted numerous failures: the tumors decreased, but often muscular reactions to recovered sexual excitations occurred—e.g., genital spasms—and, though life was often prolonged for six months to a year, some patients died. The results of cancer irradiation up until 1943 can be summarized as follows:

> Thirteen cases of cancer diagnosed at hospitals and previously treated with X rays, and 2 cases diagnosed by myself were treated and observed until 1943. All were in advanced stages of cancer cachexia. In all of the cases, the pain was greatly alleviated and the use of morphine preparations reduced or eliminated. In all cases, a decrease in the size of the tumors and an improvement of the general condition was observed. Breast tumors disappeared in all cases. . . . In most of the cases, destroyed tumor substance was eliminated. In 3 cases the orgone therapy did not prolong life. In 6 cases, it prolonged life by about 5 to 12 months and made the last few months of life much more tolerable. In 6 cases, the process of shrinking was stopped. In 6 cases, patients became able to work again. Five of the inoperable cases, otherwise doomed to an early death, are alive and in good, or at least tolerably good, condition two years after treatment. [Ibid., pp. 281–82]

When Reich first interpreted cancer as a psychosomatic disease and treated cancer patients with the accumulator, among both doctors and the public the idea of cancer being psychosomatic was so new as to be frightening or laughable. Dr. Flanders Dunbar's book on psychosomatic medicine published around this time was a pioneer study, but it drew the line at certain organic diseases. Multiple sclerosis and cancer were not considered psychogenetic in any sense. It was not until the late fifties that any number of researchers looked into such an approach to this disease. While a connection between emotions and/or life style and cancer was drawn in several pioneer articles that appeared in the reputable journal *Psychosomatic Medicine* in the late 40's and 50's, these were generally ignored for years. Other early articles concentrated on the effect of emotional stress or tension on the progress of a cancer under observation, or they studied "spontaneous remissions."

In 1959, Dr. Eugene P. Pendergrass, President of the American Cancer Society, argued in a speech to the Society that it was time to look at the relation between the emotions and cancer. "It is my sincere hope," he said, "that we can widen the quest to include the distinct possibility that within one's mind is a power capable of exerting forces which can enhance or inhibit the progress of this disease." Around the same time Karl Menninger said, "One of these days the cancer research people . . . will wake up to the fact that psychology has an influence on tissue cells, a proposition they have consistently regarded even until now as preposterous heresy" (T. F. James, p. 40).

In the fifties, a psychologist named Bruno Klopfer reportedly was able to predict with great accuracy, using Rorschach tests, the rate of cancer growth in patients whom he had never seen. Dr. Klopfer explained his success in terms of what he called "personality organization." Some people are well adjusted. Between their unconscious desires and their ego (the personality which they present to the world) there is a minimum of conflict. Others lack this fortunate balance. They may present a smiling face to the world, but deep down there may lurk, in a woman (for instance), a boiling hatred of men, or, in a man, a bitter conflict with authority. These neurotics must expend a great deal of what Klopfer calls "life energy" in keeping these emotions beneath the surface. A third type of person, the psychotic, has these conflicts, but does not have the personality strength of the neurotic to present a smiling or acceptable face to the world. Instead, the psychotic retreats from reality. He permits his antisocial feelings to erupt without restraint. Both the well-balanced individual and the extremely unbalanced individual have very slow cancer-growth rates, Klopfer found. But the person in conflict, who has trained himself to repress his real emotions, has a very fast growth rate. Why? Because, Klopfer said, he is using up a large portion of his "life energy" to keep his emotions in check.

Beginning with some studies in the late-nineteenth century a suggestive emotional theory of cancer causation linked the onset of the disease to the loss of a significant emotional relationship with a close friend, relative or spouse. "Kowal, in reviewing the literature of the eighteenth and nineteenth centuries, has shown that in numerous instances physicians have noted that a condition of despair resulting from the loss of a husband or child was the precursor of the cancerous condition" (Bakan, *The Duality of Human Existence*, p. 190). Thus, in 1931 the French cancer specialist Foque declared on the basis of his clinical experience that it was vital to study "the role of sad emotions as activating and secondary causes in the activation of certain human cancers." Foque believed that these emotions made the body's cells susceptible to the growth of cancer. He wrote, "How many times have I heard, with variations . . . the litany: 'Since the death of my child, Doctor, I am not the same . . . I cannot find my equilibrium, and that certainly is the beginning of my illness' " (*ibid.*).

So impressed were a few modern researchers—such as Dr. Lawrence LeShan, of the Institute of Applied Biology—by these early observers that they decided to study the American population, to see whether the incidence of cancer could be correlated with the loss of a strong emotional relationship. "We predicted," says Dr. LeShan, "that the age-adjusted cancer mortality rates (statistics adjusted so as to eliminate age as a factor) would be highest among the widowed, next among the divorced, next among the married, and lowest in the single group. A survey of cancer statistics showed that *in all cases where adequate statistics had been published, the predictions were borne out"* (*ibid.*, p. 41).

Other statistical studies were even more thorough. Dr. Sidney Peller was able to demonstrate that widows in all age groups had a higher cancer mortality rate than married women. He was able to rule out genetic predisposition, motherhood and social class as important factors in this difference. Studies of leukemia showed a particularly strong association between loss and depression, and the onset of the disease. In one group, seventeen out of twenty patients reported the loss of a parent or parent figure shortly before the appearance of leukemia. Even in children, the trend held true. Twenty-one out of twenty-three children experienced what could be interpreted as a separation or loss.

An impressive study of significant object loss as the emotional basis of cancer was carried out by Dr. LeShan and Dr. R. E. Worthington. They subjected 250 cancer patients to intensive interviews and tests, and found a consistent life-history pattern in 62 percent of their group, compared to 10 percent in a control group.

> [The cancer patient's] pattern included a childhood experience which made intense interpersonal relationships appear difficult and dangerous. When a strong, meaningful relationship was found and accepted, a tremendous amount of psychic and physical energy was poured into it. This . . . became the center of the individual's life. . . . He functioned adequately but superficially in other life areas. Then, for reasons beyond the individual's control, this relationship was lost. This occurred through such events as the death of a spouse, children growing up and moving away, retirement from a job, and others. Attempts to find substitute relationships failed, and the patient underwent a period of intense despair which was later repressed. Within a period of six months to eight years after the loss of the relationship, the first signs of malignancy were noted. [*Ibid.*, p. 42]

Bakan quotes a variety of other studies by responsible researchers; all of them confirm that the loss of a person close to the patient occurred not long before the onset of cancer (*ibid.*, p. 191).

Attempts to validate this theoretical approach by experiments on small animals have been carried out by a variety of researchers in the sixties. Summarizing the findings of such animal research, an article in *Psychosomatic Medicine* (May-June, 1970) concludes:

The empirical evidence reviewed in this paper justifies the conclusion that under certain circumstances, responses to cancer (tumorigenesis and mortality rates) in animals can be influenced by experimental and/or environmental manipulations of the organism. If one holds such manipulations to be psychologic or behavioral variables, it appears that there is now enough scientific justification and support for continued research efforts by psychologists interested in the responses of pathologic states such as cancer to various experiential manipulations.

Toward the end of the sixties, twenty-five years after Reich outlined his theory of cancer, research findings and interest in a psychogenic basis of cancer were beginning to accumulate. A number of investigators besides LeShan—doctors like C. B. Bahnson, David Kissen, Graham Bennette, George Solomon and psychoanalyst Godhardt Booth—were leading the way. These men held a conference May 20–22, 1968, to which they invited a variety of regular cancer research experts, including Dr. Jonas Salk. The results of these meetings published by the New York Academy of Sciences in 1969 point to a growing consensus of findings linking cancer to specific personality types and to emotional setbacks leading to depression and feelings of hopelessness.

Kissen, one of the revered leaders of the group, who devoted years to researching lung cancer, on the basis of clinical procedures with nine hundred patients had characterized them as having "poor outlets for emotional discharge." Other workers made similar observations independently at about the same time (C. B. Bahnson, p. 313). After noting intimate relationships between patients contracting lung cancer and those getting peptic ulcer, rheumatism, dermatitis, and certain neuroses, Kissen "arrived at the conviction that a [broad] psychosomatic disorder group exists, that it includes several psychosomatic conditions and that it is characterized by syn-

drome shifts from one disease to another in the same patients"
(*ibid.*, p. 314). This conviction arose from clinical studies
with over seven hundred patients. It ties in closely with
Reich's basic claim for the existence of a biopathic personality
structure. Others at the conference, including Graham Bennette of the British Medical Council, advanced a view similar
to Kissen's (Bennette, p. 361). In another paper, George Solomon quoted extensively a review study by Moos of over five
thousand patients with rheumatoid arthritis that indicates
their similarity in personality structure to cancer patients.
Thus,

> investigators agreed that arthritics, when compared with
> various control groups, tend to be self-sacrificing, masochistic,
> conforming, self-conscious, shy, inhibited, perfectionistic, and
> interested in sports. Moos and I found that female rheumatoid
> patients were nervous, tense, worried, moody, depressed, con-
> cerned with the rejection they perceived from their mothers
> and the strictness they perceived from their fathers, and
> showed denial and inhibition of the expression of anger in
> contrast to their healthy sisters. Women with definite or
> classical rheumatoid arthritis scored higher than healthy fe-
> male family members on Minnesota Multiphasic Personality
> Inventory scales reflecting inhibition of anger, anxiety, depres-
> sion, compliance-subservience, conservatism, security-seeking,
> shyness, and introversion . . . Arthritics who did poorly [in
> terms of improvement] were more anxious and depressed, were
> more isolated, introverted, and alienated, and were more un-
> able to maintain compulsive defenses and suppression of anger
> than were those with a more benign course. These findings are
> quite analogous to those already cited in the cases of infectious
> disease and cancer. [Solomon, p. 337]

In another study, further confirming Reich, Kissen "found
lung cancer patients, according to their own reporting, operate
on lower levels of autonomic activity on all . . . items [of a
specially constructed scale]" (Bahnson, p. 315). Further researches discovered that lung-cancer patients were more given
to emotional inhibition than controls, and reported significantly more adverse life conditions, both in childhood and
adulthood—for example, unhappy home, death of a parent—

than controls. "In adult life, the adverse events relate to work and interpersonal difficulties, particularly marital strife of long standing" (*ibid.*, p. 316). In some twenty-five articles in reputable medical journals between 1958 and 1968 (when he died), Kissen staked out the field, and inadvertently and unknowingly confirmed most of Reich's basic theory on cancer.

Bahnson, another serious researcher, notes at the same conference that "several workers have observed that object loss, despair, depression and hopelessness often are precursors to clinical cancer" (*ibid.*, p. 320). LeShan and Worthington discovered through psychotherapy of cancer patients a life style marked by "desertion, loneliness, and often guilt and self-condemnation" (*ibid.*). Bahnson noted that the typical "cancer patients had a primitive but unsatisfying relationship with their parents, particularly to the mother, often with lots of rage and ambivalence. Their adult relationship with intimates was often marred by distrust; if it broke down, there was little hope of warmth or solace in the environment. Typically, these people then turned in on themselves, rather than blame society, and meanwhile maintained impeccable social fronts." Further papers in this report showed that "cancer patients repress and deny unpleasant affects such as anxiety, hostility or guilt to a greater degree than do normal control subjects" (*ibid.*, pp. 322–23).

Further supporting Reich's theory, Bahnson and others at this conference presented the view that psychological or somatic disease can be seen as complementary. Reich emphasized this in the case of schizophrenia and cancer. Bahnson uses a psychophysiological model "to conceptualize psychological and somatic diseases as alternative or complementary results of conflict and repression . . . both malignancies . . . and deteriorated psychoses . . . represent the most extreme forms of disorganization and self-alienation" (*ibid.*, p. 324). As others have done, Bahnson points to numerous studies linking cancer to changed hormone production—which links it to the sexual area. He notes two studies with rats, showing that "those prematurely separated from their mothers have greater mortality rates from implanted tumors, [while] . . . rats handled and petted early in life resisted Walker Carcinoma 256 longer than rats raised in isolation and without handling" (*ibid.*, p. 329). In short, experiments with animals prove that early and regular affection and satisfying interrelationships can be immunizing against cancer.

Dr. Stephen Black, eminent British doctor and writer, in his *Mind and Body*, refers to the work of Professor Tromp, who postulated emotional factors in cancer in an article in 1965.

Tromp points out that Andervont (1944) and Muhlbock (1950) demonstrated in mice that community size, the type of housing and degree of activity of the animals does . . . affect the incidence of mammary tumours. Working with strains particularly liable to mammary carcinoma, they found that of 50 male and female mice, in one large zinc cage, 29% of the females developed mammary carcinoma. But of 50 mice in 10 zinc cages of 5 mice each, the proportion was nearly double at 56%; while distributed in the same way in glass cages, the proportion rose to 67%. When each animal was kept alone in a glass cage the proportion was 83%. [Black, p. 114]

Throughout the 1968 conference various insights and observations were constantly repeated and refined (Bennette, pp. 358–59). For example, a paper by Katz *et al.* concluded that psychological variables may play a role in breast cancer (Katz, p. 515); an elaborate study of M. and C. Bahnson shows cancer patients to be low in projection of anxiety, depression, guilt, hostility and dominance (Bahnson and Bahnson, p. 550). Their denial of unpleasantness to the external world is a significant indicator of repression. Booth notes that "epidemiological studies indicate that the primary tumor develops in the organ related to the frustrated psychophysiological relationship" (Booth, "General and Organic-Specific Object Relationships in Cancer," *Annals of New York Academy of Sciences*, Vol. 164, Art. 2, p. 572)—for example, the breast in women who need, but fail to receive, the pleasures of being sucked or touched. Little is said in this thick volume about any sexual basis for cancer, and one gets the impression that most of the investigators were loath to raise such intimate questions with their patients. But implicitly, if not explicitly, it is clear that erotic blockages were involved.

Significantly, research by Ruth D. Abrams and Jacob E.

Finesinger into the attitude of cancer patients, carried out by extensive interviews, concluded: "The most significant and characteristic concept held by our patients was that cancer was a disease of unclean origin. . . . The idea that cancer is a 'dirty disease,' 'unclean,' 'repellent' was repeated over and over again" (Abrams and Finesinger, "Guilt Reactions in Patients with Cancer," *Cancer*, Vol. VI [1953], p. 478). Sexual demands by spouses were frequently cited as a cause of cancer.

Some sexual malfunctioning seems, from certain studies, to be associated with the development of cancer. Thus, A. Beatrix Cobb, in an unpublished doctoral dissertation, "A Social Psychological Study of the Cancer Patient" (University of Texas, 1953), observed that in men with cancer of the prostate, there was an unusual "sexual preoccupation leading to multiple marriages" (p. 52). Milton Tarlau and Irwin Smalheiser, studying women with cancer of the breast and cancer of the cervix, concluded that there was "a general disturbance in sexual functioning" in both groups, their attitudes toward sexuality entailing rejection of the feminine role and "uniformly negative feelings toward heterosexual relations" ("Personality Patterns in Patients with Malignant Tumors of the Breast and Cervix: An Exploratory Study," *Psychosomatic Medicine*, Vol. XIII, No. 2 [1951], p. 118). It was found that these women characteristically had mothers who had warned them to stay away from men, and that their reaction to the onset of menstruation "was uniformly one of rejection, ranging from feelings of fear, shame and disgust to strong hysterical outbursts" (*ibid.*). The Tarlau and Smalheiser study, which was the first to make this observation, suffered methodologically, especially in the lack of a control group for comparison. It was replicated by John I. Wheeler and Bettye McD. Caldwell, with the addition of a control group. They confirmed the conclusions of Tarlau and Smalheiser, finding in both breast- and cervical-cancer patients "greater negative feelings toward sexual relations" ("Psychological Evaluation of Women with Cancer of the Breast and of the Cervix," *Psychosomatic Medicine*, Vol. XVII, No. 4 [1955], p. 264). Catherine L. Bacon, Richard Renneker and Max Cutler, in a detailed study of the personalities and personal histories of forty women with breast cancer, similarly concluded that "sexual inhibition and frustration" were considerably higher than what "we normally observe in our clinical investigations of neurotic women" ("A Psychosomatic Survey of Cancer of the Breast," *Psychosomatic Medicine,* Vol. XIV, No. 6 [1952], p. 455).

A comprehensive psychoanalytic study of five cases of breast cancer underlines the relevance of these associations. It concluded:

> That women with cancer of the breast tend to accept lovers or husbands who are extremely unsatisfactory. These women experience . . . frustration of feminine needs through the choice of an inadequate type of lover or husband. The mates of our patients were outstandingly unsatisfying in any of several ways: they were cold, sadistic, alcoholic, seclusive, impotent, uninterested, opposed to having children, or monumentally narcissistic. [Bakan, p. 184]

A controlled study by Reznikoff of women with cancer compared with women with benign tumors or no tumors revealed that "in addition to viewing men as not gratifying their needs for affection and less frequently displaying contentment with their relations in this sphere, the cancer subjects, compared with the normal women, on fewer occasions conceived of men as protective or sympathetic" (Marvin Reznikoff, "Psychological Factors in Breast Cancer: A Preliminary Study of Some Personality Trends in Patients with Cancer of the Breast," *Psychosomatic Medicine,* Vol. XVII, No. 2 [1955], p. 102). Another finding—that "cancer subjects were more ambivalent toward accepting responsibilities associated with raising children and distinctly more fearful and threatened by pregnancy and the birth process."

Studies of men have also indicated that males with cancer of the prostate are superficially very "nice," in the sense of not expressing aggression and being very compliant and superficially cooperative (Cobb, *op. cit.,* p. 39). In twenty-five of their forty cases of breast cancer, Bacon, Renneker and Cutler observed that there was . . . excessive pleasantness . . . under all conditions and a common inability to deal appropriately with anger. Thirty had no technique for discharging anger directly or in a sublimated fashion. Most of these even denied having ever been angry. These were the ones who maintained a cheerful, pleasant façade through all adversities. Friends were prone to describe them as "the nicest woman we know," "she wouldn't hurt a fly," or "always thinking of others."

In Britain, during the sixties, two doctors initiated attacks on orthodox medicine's approach to cancer. First, "the research that won Lang Stevenson the Lawson Tait memorial prize in 1962 was based on the biological hypothesis that cancer is a stress reaction: genetics and other predisposing causes may determine the nature of our diseases, he wrote, but 'it is defensive reaction to an abnormal situation that initiates the change' " (B. Inglis, p. 262). Then Arthur Guirdham's *Cosmic Factors in Disease* (London: Duckworth, 1963) was another onslaught on orthodoxy, with its plea for a re-evaluation of what illness is for. "The stress disorders, Guirdham argued, represent a protest, a message, and a mask: the protest, registered by the individual's real self against his conditioned, environmental personality; the message, an indication what is the matter; but the mask, a way of excusing himself for his predicament, an intimation of helplessness, as if the individual hopes by becoming a patient to be absolved from responsibility" (*ibid.*, p. 263).

Where such psychosomatic cancer research is admittedly weakest (and this is partly shared by Reich's work) is in clear evidence of the physiological mechanism or process by which emotional states lead to an actual cancer. Reich claimed that it is due to sequestering or blocking of orgone flow, leading to putrefaction. No orthodox cancer researcher to date seems to have examined the orgone theory, nor Reich's claims for the existence of T-bacilli. Till this is done, we can only conclude that in rough outline Reich's cancer biopathic theory is being slowly confirmed, although, on details, it may later be found seriously wanting.

With diseases other than cancer, Reich found the accumulator to be more effective. He told of anemias being eliminated in three to six weeks of daily irradiation; he claimed the disposition to colds is greatly reduced; blood pressure is lowered in cases of vascular hypertension; in cases of angina pectoris, either the subject's condition is improved or the symptoms disappear altogether; constipation is reduced; and even arthritis seems to be greatly improved. With the use of an orgone shooter, a device to concentrate the energy on a local body part, burns, wounds and bedsores heal very rapidly. Ordinary ulcers are healed, and even a case of varicose ulcers yielded to the application of the shooter in six weeks. In this case and in some others, however, the complaint returned and necessitated further orgone irradiation.

In the middle and late forties Reich collected around himself a number of colleagues, including some twenty doctors and psychiatrists, and they began to apply the orgone. From time to time they described their results in journals like the *Orgone Energy Bulletin (OEB)*. Thus Dr. Simeon Trapp reported in "Orgone Therapy of an Early Breast Cancer" (*OEB*, July, 1950, pp. 131–38) how several breast cancers totally disappeared in about six weeks. In the January, 1951, *OEB*, Dr. Emmanuel Levine reported on "Treatment of a Hypertensive Biopathy with Orgone Energy" (*OEB*, January, 1951, pp. 23–34). In an elaborate medical and psychological history he recounted how the blood pressure fell and the patient came to feel much better, and how relapses occurred a couple of times but after two years all signs of the complaint had disappeared.

A physician who made the most extensive use of accumulators during this period was Dr. Walter Hoppe, of Tel Aviv, Israel. He went from five- to twenty-layer accumulators and described his results in an article, "My Experiences with the Orgone Accumulator," in the *Orgone Energy Bulletin* (January, 1949). He described a wound treatment, which the surgeon in a review of the case called "a miracle cure." He mentioned successes with constipation, angina pectoris, arteriosclerotic heart disease, chronic bronchitis, chronic diarrhea and a gastric ulcer. Unusual complaints like inflammation of the eyeball, vasomotor disturbances, paradentosis (bleaching of the gums), and eczema, all yielded after four to eight weeks of daily irradiation. Some details on the last-mentioned case are interesting:

When [the patient was] 31 years old . . . the first indications of the present osteoporosis (stiffening of the right femur) were visible after a cramp of the calf. The patient complained of pains in her knee, which constantly increased. An X-ray picture supported the diagnosis of osteoporosis. . . . X-ray depth radiation was unsuccessful. Further treatment was given up. The right knee could be bent only with severe pain. The pains fluctuated in intensity, but the patient was never free from pain. In this condition, she came for orgone treatment. After 14 days of using the twentyfold accumulator, the pains had completely disappeared and, for the first time in years, the patient was able to bend her knee completely without feeling any pain. Instead of these ailments, a thrombosis formed in the *left* leg which forced the patient to use her right leg to stand on, which did not bother her in spite of considerable physical exertion . . . [*Orgone Energy Bulletin,* January 1949, pp. 19–20]

Dr. Hoppe's article confirms Reich's findings regarding the prickling sensation commonly felt in the accumulator, the heat generated and the rise in body temperature. He added:

"Suggestion" was out of the question here, because many patients felt the prickling sensation without knowing that this was a characteristic reaction in the accumulator. This prickling sensation shows up irregularly and with varying intensity. Another striking phenomenon is the fact that a patient's face is often reddened when he leaves the accumulator. [*Ibid.,* pp. 21–22]

Since he is one of the few investigators to use a powerful twenty-layer accumulator, his findings are worth noting. Hoppe explained that

the difference between treatments in the twofold and the twentyfold accumulator lies in the quicker and more intensive efficiency of the healing process in the twentyfold. In many cases, especially in the twentyfold accumulator, a feeling of pressure in the head appears, in some cases nausea. We had a woman patient who regularly, after 5 to 7 minutes had to leave the accumulator because of nausea. In a single case, perspiration broke out together with an acute feeling of weakness, which, however, disappeared soon after the patient left the accumulator. The majority of patients feel rested and relaxed after sitting in the accumulator. Signs of fatigue disappear. [*Ibid.,* p. 22]

Another physician, Dr. N. Wevrich, experimented with the accumulator on a number of unusual conditions, including diabetes. His patients showed cessation of pain from a scalding with hot water, after fifteen minutes' irradiation by an orgone shooter, overcoming of recurring fatigue after a quarter hour in an accumulator, and eradication of insomnia. In the case of the diabetic, using the accumulator regularly made a normally tense person more relaxed and less apprehensive, less inclined to take additional insulin, and more able to cope with sudden outbreaks of emotional stress (*ibid.,* April, 1951, pp. 110–12).

Owing to its considerable healing effects, the accumulator came into greater demand after 1946. An article by Mrs. Ilse Ollendorff Reich in the January, 1951, issue of the *Orgone Energy Bulletin* reports on the chronology and financial aspect of accumulator usage up to 1950. By 1943, some twenty had been constructed, mostly onefold types. Those renting them paid ten dollars monthly to the Orgone Energy Research Laboratories; by 1945, fifty had been built. In the next two years, about fifty accumulators per year were ordered; these were of a two-layer variety. This figure jumped to nearly one hundred per year in 1948 and 1949—these were mainly threefold, with a few five-layer ones. By the end of 1950, 322 accumulators were in use in the United States.

Twenty-seven of these accumulators are given out free of charge, 11 accumulators built by private people for their own use, and 20 accumulators are used at a reduced rate. All of these accumulators are still given out on an experimental basis. Every user has to sign an affidavit to that effect, and each affidavit has to be signed and the diagnosis filled in by either the treating medical orgonomist or, in cases where the patient is geographically too far removed, the affidavit has to be accompanied by a statement on the physical status of the patient by his own physician. [*Ibid.,* January, 1951, p. 54]

In the New York area, Dr. Simeon Trapp ran an Orgone Energy Clinic, in which free treatments with the accumulator were available. By 1950 it was estimated that probably a thousand people had made some use of an accumulator, since in many cases family members and friends availed themselves of it from time to time.

While the income from the accumulators rose from "$4,594 in 1946 to $23,000 in 1950" (*ibid.*, p. 55), the total costs of construction amounted to $20,400. Surpluses went to the lab, whose annual expenses for salaries, overhead and equipment rose from $9,500 in 1946 to $21,800 in 1950 (*ibid.*, p. 56). Research expenditures not met by income from the accumulator rental were secured from contributions of co-workers, gifts by Reich, general donations and a lecture tour by A. S. Neill, founder of Summerhill. In short, accumulator rental provided only a part of the costs of running the lab. Reich took nothing for himself personally. The whole operation up to 1950, was obviously quite unimpressive quantitatively, yet at this very time the Food and Drug Administration, apparently egged on by some doctors and psychiatrists, were assiduously preparing a legal case against Reich. The charge—fraudulently advertising the accumulator as a cancer cure!

Little independent checking on the accumulator's powers went on in the fifties. Practically nothing was published of scientific tests except by Reich's co-workers. Private individuals here and there built accumulators, but few kept systematic records of results.

The only one of such investigations to come to my attention was from a small organization in Kenmore (near Buffalo), New York, by Tech Products. This research team first read Reich in 1953 and, after a while, built a three-layer blanket and began to use it daily for fifteen minutes to an hour, making careful checks on sensations and objective changes. Here are excerpts from a report mailed to me in the late fifties. (The investigators were not members of Reich's organization, nor known to me personally.)

> The very first reaction noticed in the subject was an almost immediate light sweat, although the temperature in the accumulator was never over 75° F., and the subject remained quiet and relaxed. Next, it was noticed that the subject showed all the symptoms of a "cold" each time he entered the accumulator. This condition prevailed for approximately ten sit-

tings. It is worthy of notice that the stuffed nose and runny eyes cleared up a few minutes after the subject left the accumulator. After these symptoms failed to appear again (as stated, at the end of ten sessions) the subject's back became red at each session, finally retaining a dark, flushed appearance, as of sunburn. During this period the subject noticed a feeling of extreme well-being and good health.

An interesting point is that the two house cats would go to any lengths to enter the accumulator and curl up for hours. It made no difference if the door was open or closed, or if the light was on or off.

An extremely interesting case concerned a woman (age 37) who on examination by her doctor was found to have a small growth in the uterus . . . The growth stabilized at a particular size for about a year, and the doctor then began to consider surgery as necessary. A three-ply orgone blanket was made, and the woman used it faithfully as a covering at night. The next three-month examination completely flabbergasted the doctor, as the growth had disappeared! A year later has brought no reappearance.

Other cases of effective use of blankets have been: rapid regression of all symptoms of arthritis in hands, arms, and shoulder of a 27-year-old woman; removal of arthritic symptoms in Atlas region of 55-year-old woman; removal of very large wart of nine years' standing after one fifteen-minute application of "Orgone" energy from a "shooter" accumulator . . .

We made a small (one cubic foot) three-ply accumulator, and allowed it to "charge" for about a month. A package of hybrid Bantam corn seeds were split into two parts. One of the portions was placed in the accumulator, and the other was wrapped in a paper bag and stored on a shelf. This was done in the first week in October, and storage of both portions was in an unheated barn. In May, the seeds were planted in side-by-side rows, in carefully measured holes three inches apart, in marked rows.

We did not know what to expect, as no work has been done in agriculture along such lines . . . as far as we know. Consequently, we were greatly surprised when we received—as far as we could tell by count—100% germination in the "charged" rows, and the usual or "normal" 80% germination in the "un-charged" rows. Earth and water conditions were as close to the same as possible.

Further experiments in agriculture consisted in the enclosing of tomato plants in individual accumulators open at the top and bottom. These accumulators are simply food cans with the tops and bottoms cut out, and the paper covering left on. The can is slid over the plant and down, and is pressed about one inch into the earth . . . Twenty-four plants were set out in this fashion, along with twenty-four controls from the same plant. The difference was quite remarkable; much heavier trunks and far lusher foliage, with more tomatoes of the same size, taste, etc. . . .

The same series of tests on the corn and tomatoes were run during the following two seasons, and results were very similar. [Handwritten report, Tech Products]

Further application of orgone to agriculture is discussed briefly in O. Raknes' study *Wilhelm Reich and Orgonomy*. Raknes notes that "seeds of different kinds that have been kept . . . in orgone accumulators before being sown gave more abundant crops and in shorter time than seeds of the same kinds that had not been so irradiated" (p. 109). He also mentions a researcher who irradiated sickening plants and watched them grow healthy and lush again.

An interesting and easily reproducible experiment was reported by Barry Sheldon, in *Orgonomic Functionalism*, the British neo-Reichian journal, in 1964. Using Dunn's curled water cress, the experimenter watered his plants daily, the one with water being kept in a small one-cubic-foot orgone accumulator, the other with water kept in a similar-size control box. The seeds were kept in trays placed on slates, left inside their respective boxes, so the experimental seeds received orgone water and orgone energy directly from its accumulator. After about two weeks he reported that "the orgone-irradiated seeds germinated much more quickly and in greater numbers than did those in the control box. Orgone-irradiated seedlings progressed at twice the rate of the others: they display a greater strength of stem and a more extensive root development. [They] also smelled sweeter" (*Orgonomic Func-*

*tionalism,* Vol. 8, No. 1, p. 61). Clearly, there is fertile ground here for simple and effective testing of the orgone, which if it proves to work will yield great benefits to agriculture.

While evidence for orgone's effect on plant growth is very limited, the survey of medical research suggests that at least in broad outline, Reich's theory of biopathic disease is not to be lightly dismissed. Clearly, emotional and biological energies are closely tied in to the health or breakdown of organisms. Cellular structures and processes are basic. Whether Reich's explicit formulation of the biopathy is correct or not, it seems to offer some pretty shrewd insights into many degenerative-disease processes.

*Chapter Eleven*

# ORGONE AND WEATHER RESEARCH

The fertile imagination of Wilhelm Reich led him in the late forties to a variety of ingenious, if not improbable, experiments to tap further uses for orgone. Some of these, like the attempt to nullify atomic radiation through orgone concentration, proved dangerous and were not repeated by followers or supporters. Others, like weather control and rain making, have been subsequently repeated with apparent confirmation. Throughout this period of restless search for new applications of the orgone, Reich either left behind some of his former friends (e.g., A. S. Neill) or antagonized and frightened others away. Few could really tolerate the tremendous pace and impatience (for discovery) which animated him. On many occasions he was asked why he could not slow down and take time to confirm more thoroughly his "new" discoveries, rather than to leap on to others. His usual answer was that his role was to uncover new implications and uses for the orgone; let others pursue the more pedestrian task of precise validation and replication! "My job is discovery, and I leave it to others to carry out the results" (Neill, Ritter, Sharaf and Waal, p. 28).

One of the more abortive applications of orgone was to produce a sufficient concentration to turn a motor. It appears from the literature that neither Reich nor any of his close associates took time to follow up this invention. It began when

Reich purchased in May, 1947, a Geiger-Müller Portable Field Set designed to determine gamma rays, X rays and cosmic rays. It was placed in the orgone accumulator. After giving off several clicks per minute—a background radiation count—it appeared to go dead and for about a week no such clicks were emitted. Then, three and a half months later, Reich gave it a fresh try only to find it producing, according to the pointer, six thousand clicks per minute. Tests showed these originated from the counter tube which had apparently become saturated with orgone by exposure in the small three-layer accumulator. Compared to "normal" clicks from atomic radiation, this was extraordinarily powerful.

A few months later, Reich, utilizing vacuum tubes thoroughly soaked in orgone, registered impulses (clicks) of up to 100,000 per minute, turning the counter pointer around 1,000 times in a minute. Reich theorized that here was a new motor force, since no ionization effects could be this powerful. The next step was to link this force to a motor. On June 24, 1948, in the presence of five witnesses (members of the Orgone Research Lab staff) Reich purported to have set a motor (Western Electric KS9154 Serial No. 7227) in motion, using an activated filament of electronic amplifiers (*OEB*, January, 1949, p. 11). He further claimed that the speed of this motor could be regulated and that the potential varied, as natural with orgonotic functions, with such phenomena as the weather and time of day.

One "ingredient" in the apparatus was kept secret and labeled in the article describing the experiment, simply as Y. Speaking in February, 1971, with psychologist Dr. Myron Sharaf, one of the five witnesses of this test, I was given to believe, it was a small electric battery. How this fact would affect a scientific evaluation of this "new" force I am unable to say. The comments on this motor by A. S. Neill, of Summerhill, are interesting. "Ten years ago in Maine I saw a small motor turning over when attached to an orgone accumulator. 'The power of the future,' cried Reich joyfully. But as far as I ever knew, the experiment was not continued" (Neill, Ritter, Sharaf and Waal, p. 28).

In the autumn of 1950 Reich, concerned over the spread of nuclear radiation and its potential dangers in an atomic war, decided to see if the orgone could control this harmful radiation. In various experiments, he thought he noticed an antagonism or basic difference between radioactivity and the orgone, and he considered it possible that orgone might constitute a defense or shield against radioactivity. So he purchased some radioactive isotopes from the Atomic Energy Commission and put them into the large accumulator in his lab at Rangeley, Maine. His expectations that the orgone would weaken or control the radioactivity backfired when a very different series of events resulted. They are written up in detail in *The Oranur Experiment*.

Fundamentally, what apparently occurred was that orgone excited the isotopes and spread their dangerous radiation far from their enclosure in the accumulator. Some Reichians give a different version of the interaction, but all agree on the physical effects. For example, Raknes says: "The radioactivity triggered, or 'excited,' or 'stimulated' the orgone, so that all those who took part in the experiment were taken ill, even though they had strictly observed the precautionary measures prescribed by the Atomic Energy Commission" (Raknes, p. 33). Either "excited" orgone or more potent radioactivity resulted, so that symptoms of radiation sickness occurred and one worker nearly died. Pieces of limestone rock for several hundred yards on all sides of the lab turned a peculiar brownish or blackish color and seemed to be disintegrating. This I observed personally on my visit to Rangeley in 1953. The building in which the isotopes were "housed" and the grounds around for some distance became radioactive and could not be safely used for almost two years. Raknes tells us that even short visits to the infected area could lead to dizziness and fainting and might have been fatal. "Strongly armored persons did not feel (at first) or at least only to a slight degree (its) effects . . . and would (later) faint and become helpless, without having felt any warning symptoms" (*ibid.*, p. 77). The emergency, which lasted from March to September, 1952, became a major crisis for the small organization. Several of Reich's workers and collaborators became upset or afraid and withdrew from the work. Others naturally asked: "How had Reich miscalculated so badly?"

I am unable to follow the technical aspects of the Oranur

experiment, nor am I able to evaluate critically what it implied in detail regarding the orgone energy.* But it seems clear that Reich definitely miscalculated the power of the orgone to control or nullify radioactivity. He jumped too quickly to conclusions about the interaction of the two energies, due in part, no doubt, to his impatience to be first to discover new things, but also to his passion to help mankind. However, what is incontrovertible is that a dangerous radiation ran wild around the lab and affected people, rocks and other natural phenomena. Whether this was caused by the interaction of orgone and radioactivity, by the careless storing of the isotopes, or by some combination of radium and other substances, it seems well-nigh impossible to say at this time.

A side effect to the radioactivity emergency, however, led to further Reichian inventiveness. For the first time since arriving at Rangeley in 1945, Reich and his co-workers observed in April and May, 1952, strange weather conditions, subsequently labeled "DOR" (deadly orgone radiation). The salient visible effect was peculiar "black" clouds that drifted over the area and seemed to get stuck there. Reich felt they were the result of the interaction of the orgone with the radioactive isotopes. Did radioactivity transform the orgone, making it something evil? Here, summarized from the October, 1952, *Orgone Energy Bulletin,* is a description of these strange weather conditions—and their correlation with very peculiar Geiger-counter phenomena.

A "stillness" and "bleakness" spread over the landscape, rather well delineated against unaffected surrounding regions. The stillness is expressed in a real cessation of life expressions in the atmosphere. The birds stop singing, the frogs stop croaking . . . The birds fly low or hide in the trees. Animals crawl over the ground with greatly reduced motility. The leaves of the trees and the needles of the evergreens look very "sad"; they droop, lose turgor and erectility. Every bit of sparkle or luster disappears from the lakes and the air. The trees look black, as though dying. The impression is actually that of blackness, or better, bleakness. [Pp. 171–72]

---

* Those who wish to understand the physical findings should read *The Oranur Experiment.*

Here in this beautiful piece of nature's wonderland, in remote Maine, the sparkle of life seemed to have left the landscape.

The unique characteristic of this condition was a bleak blackness. It seemed to hover strongly over swamps (where decay processes are strong) and landscapes without vegetation. The green color of trees and meadows disappeared from the mountain ranges around Rangeley. They appeared somehow "dirty," or blackish with a purple tint. The water in the lakes became calm and motionless. Thin clouds which seemed the focus for this antilife radiation would frequently appear without obscuring the sun. In other words, one could often differentiate this condition and its bleakness effects from other parts of the total area in which the sun shone and gave a sparkle to the air and a blueness to mountain ranges.

Both persons and Geiger counter showed remarkable reactions to the lifeless black clouds. The more sensitive, including some farmers of the district, labeled them "bad air" or "some atom dust." One would say "There is something wrong in the air"; another, "I cannot get any air"; or "I feel it sometimes like something closing in on my face." The insensitive often felt nothing or just mentioned more "heat" than usual. The reaction of "normal" people was that the atmosphere seemed oppressive and almost suffocating, like the ominous stillness one may feel before a violent electrical storm. Reich noted that the clouds apparently came from the west and that there had been a tornado west of Rangeley on March 21, 1952. The reactions of the Geiger-Müller counter to the overhead passage of the black clouds were variable and strange: it would race to the limit of 100,000 clicks per minute, then fall rapidly or fade to nearly zero and then race upward again. (Such high counts would cause alarm in an atomic plant.) Reich notes that the counter seemed to become nervous and unable to make up its mind whether to race, to fade or to jam. This erratic behavior was ascribed to an overcharge of the counter tube—too much energy in the area!

The exact character of these black (DOR) clouds remains a mystery. They might have been simply black thunderclouds (cumulo-nimbus) carrying a heavy electric charge whose effects on the surrounding life forms were exaggerated by Reich and his co-workers. An outstanding Dutch physicist, S. W.

Tromp, and other scientists who have investigated the field of biometeorology have noted how some people are sensitive, a day in advance, to the onset of thunderstorms or other major weather changes. Pains in the lower-back area or headaches provide them with warning signals. It seems evident that weather conditions can noticeably depress organisms, including animals and humans; perhaps the emergency at Rangeley made Reich unduly sensitive and what he labeled DOR were simply depressing, enervating weather conditions, heralded by black clouds, and that he looked unnecessarily for some singular explanation.* Or perhaps these clouds were associated with spill-over energy from typhoons or hurricanes or with atomic blasts, or noxious concentrations of urban smog. It is interesting that in the late sixties observers have noted large ominous concentrations of blackish clouds, which seem to be threatening to life, near or off the coast of Maine and other Northeastern states.

It is significant that Reich felt called upon to disperse these DOR clouds. In so doing, he invented a new device, a "cloud-buster," which can be viewed as another application of the orgone theory.

Remembering an unexplained phenomenon from 1940, when the casual pointing of long metal pipes at the surface of a lake seemed to affect the movement of the waves, Reich experimented in April, 1952, pointing long thin metal rods at the black clouds. His aim was to disperse them. So, he linked the metal pipes, 9 to 12 feet long and 1½ inches in diameter, through BX cable† to a deep well on his Rangeley property, theorizing that, as water is a great absorber of orgone, it might draw and absorb the DOR.

He pointed these pipes at the black clouds and claimed, "the effect was instantaneous: the black DOR clouds began to shrink. And when the pipes were pointed *against* the OR (orgone) energy flow—that is, toward the west—a breeze west to east would set in after a few minutes' 'draw,' as he came to call this operation; fresh, blue-gray OR energy moved in where the nauseating DOR clouds had been a short while be-

---

* S. Tromp's *Psychical Physics* and *Medical Biometeorology* are competent, scientific analyses of weather and its various psychic effects.
† Metal-covered flexible casing, with a diameter of 2½ or 3 inches.

fore. Soon we learned that rain clouds too could be influenced, increased and diminished, as well as moved, by operating these pipes in certain well-defined ways" (*OEB*, October, 1952, p. 175). The process is somewhat clarified in an article entitled "Cosmic Orgone Engineering," in which Reich says, "DOR clouds are encircled by highly excited orgone energy" (W. Reich, *Selected Writings*, p. 453) whose aim is apparently to contain and combat this evil force. As disease entities or infections in the body become surrounded by white blood cells that sequester and then kill the "intruder," so DOR in the atmosphere tends, according to Reich, to pull around it a ring of excited orgone which holds in and then tries to eliminate the dangerous radiation. He adds: "When drawing fresh orgone energy from the west or southwest, whirling air currents develop, similar to the 'dust devils'* in deserts and in regions developing into desert. The atmospheric Geiger reaction may reach 100,000 or more counts per minute as a sure sign of high excitation. . . . It is as if the atmosphere is feverish. Strong winds usually in the form of sudden gusts develop" (*ibid.*). Dust devils and tornadoes may be the visible aspects of excited orgone's activity.†

During this period Reich found that the cloudbuster could affect ordinary rain clouds. He constructed two cloudbusters in September–October, 1952, aimed at the destruction *and* creation of clouds, rain making and the stopping of rain. Thus, from a situation of "unbearable" concentrations of DOR, Reich was led to a whole new series of speculations and experiments eventually aimed at weather control. In its final form, the cloudbuster consisted of a number of hollow metallic tubes which, by a telescopic arrangement, could be lengthened or shortened. They were attached to a stand and could be turned in any direction, vertically or horizontally. The upper ends were fitted with caps so they could be closed or opened at will. The lower ends were joined to flexible BX tubing which led into water or moist earth. One must give the

---

* "Whirlwinds known as dust devils, consisting of warm air rotating rapidly and entraining fine sand, occur especially in desert areas" (S. Singer, p. 95).
† There may be a connection between tornadoes or dust devils and ball lightning (S. Singer, pp. 96–97).

man full marks for his imagination and ingenuity!

To explain how metal pipes could draw energy from cloud formations, Reich pointed to the known functioning of the lightning rod. This is a metal rod, usually mounted on a building, but not necessarily so, which attracts the lightning discharge and conducts it through heavy wires into the ground. As the charge is grounded the danger is overcome. As noted in Chapter One, Reich had contended that static electricity (lightning) is actually orgone. He now claimed:

> This lightning-rod system functions according to *orgonomic*, and not according to electrical principles: in the lightning rod system, the atmospheric charge is drawn *from* the atmosphere *toward* the point of the rod and further *toward* the earth's crust. It is, thus, the *orgonomic* potential *from weak to strong* which is operative in the case of the lightning rod. If the electrical potential from high to low were operative in the lightning rod system, the direction of flow would necessarily be the reverse, from the earth's crust toward the atmosphere; the energy would stream off and *away* from the point of the lightning rod. [*OEB*, October, 1952, p. 178]

Reich assumes that the earth has a higher charge than a lightning cloud. Most persons would take the opposite view, since the energy potential in a black storm cloud is so visible. They would accept orthodox science's view that the static electricity in the cloud follows the accepted process of moving from a high- to a low-potential area. Reich's assertion here goes against this interpretation, but fits in with various orgonotic phenomena like the temperature-difference experiments which call into question the second law of thermodynamics.

The operation of the cloudbuster, according to Reich, differs from that of the lightning rod on four points. First, its purpose is to draw orgone energy out of the atmosphere and clouds, but *slowly* and in small amounts at a time. Its action is not immediate as with a lightning rod. Second, its pipes are longer—at least four meters long—and they are hollow, not solid. (The reason for their length and hollowness is not clearly evident.) Third, their aim is to trigger a flow of atmospheric orgone energy in a specific direction. Theoretically, once the direction of the flow has been established, it will continue that way until another artificial or natural stimulus changes it. Fourth, and perhaps most significantly, the pipes draw the charge to water, preferably the *flowing* water of brooks, lakes and rivers. (Reich found that the attraction between water and orgone is very strong.)

Professional students of water divining like Cecil Maby and Professor S. W. Tromp insist that moving water—for example, in underground streams—sets up a certain force field. This is strictly a conventional electrical effect. It may be that this force field acts as a link to the electrical charge in the earth, and so creates a potential different from that in the cloud. This might be a fairly low charge, and in such a case the flow of energy would be from the highly charged cloud to the flowing water. To what extent the still water in a well would mesh dynamically with the nearby earth's electric field is another question. Significantly, later attempts at cloudbusting, in which the cloudbuster was connected to still water, did not work too well. Those neo-Reichians in the United States who tried to confirm Reich's weather experiments generally seemed to prefer running water. It would appear, therefore, that without accepting Reich's unorthodox conception of the charge flow, one *could* expect that at times the cloudbuster might draw an electric charge from cloud to running water.

Reich admitted in 1952 that much remains to be learned about cloudbusting, but he was certain that the basic principle had been discovered. Thus, he says:

> One dissipates clouds of water vapor by withdrawing, according to the orgonomic potential, atmospheric (cosmic) OR (orgone) energy from the center of the cloud. This weakens the cohesive power of the cloud: there will be *less* energy to carry the water vapors, and the cloud necessarily must dissipate. The orgonomic potential between cloud and its environment is lowered. [*Ibid.*, p. 180]

For cloud creation (and thus rain making) the same principle is used. However, the cloudbuster pipes are aimed not at the center of the cloud as for "busting" but to one side of it. This apparently acts to enlarge the cloud. The more clouds present and the heavier they are, the easier it is to produce cloud expansion and rain. Conversely, the fewer the clouds, the more difficult and the longer it takes for the clouds to give up their water. The mechanism, said Reich, in both cases is that the pipes when focused on a cloud-free section of the sky disturb "the evenness in the distribution of the atmospheric orgone energy" (*ibid.*). A concentration of orgone is produced which attracts water vapor to it and produces a new cloud.

According to Mrs. Reich's biography and other co-workers who were there in Maine, Reich succeeded in bringing rain to the Rangeley area on more than one occasion and actually received praise for this in the local papers. The text of these newspaper compliments is reprinted in relevant articles in bulletins of the Orgone Institute. But serious problems can arise through rain making. Reich noted:

> Strong reactions to cloud-busting in Rangeley, Maine, have been observed in distant regions (Boston); such influence on far-away regions is due to the continuity of the orgone envelope; the details will require extensive and careful study. We have always been cautious not to overdo cloud-busting, since small twisters and rapid changes of winds have been observed beyond any reasonable doubt. Also, on one occasion, heavy, prolonged rain occurred upon faulty operation. [*Ibid.*, pp. 180–81]

Raknes also notes that Reich was aware of how certain uses of the cloudbuster could be disadvantageous to neighboring regions and so was concerned to use this device in responsible ways.

In his book *Weather Changers*, D. S. Halacy, Jr., notes: "There are many obvious dangers in tampering with the weather, such as creating arid zones . . . and detrimentally altering the pattern of winds, or inducing tornadoes or other damaging storms." Other dangers are disrupting the plant and animal life and even human communities (pp. 196–97). Irresponsible use could result in not only displacing rain from needy areas to other locations but even upsetting the balance of the region's climatic process. To conduct rain making responsibly means finding out the weather conditions over wide areas and securing the cooperation of meteorological officials. Needless to say, Reich never succeeded in this.

He did establish a unit of his work called Cosmic Orgone Engineering (CORE), but it never received much of his time,
mainly because of his forced preoccupation with court proceedings instituted against him at about this time by the
Food and Drug Administration.

An account that psychiatrist Dr. Ola Raknes, Reich's chief
follower in Norway, gives of cloudbusting at Rangeley is
rather valuable:

> One day Reich was experimenting with the cloudbuster, trying
> to find out in which direction it would be most profitable to
> point the apparatus. While he was pointing it in different
> directions I happened to notice that the wind, a light breeze,
> over the nearby Dodge Pond, was changing its direction. Not
> far from the cloudbuster was a weathervane, and I decided to
> follow its movements. Whenever Reich changed the direction
> of the cloudbuster, the weathervane would show in a few
> minutes that the wind had changed correspondingly. I was
> strongly impressed by this observation.
>
> A couple of days later Reich, who was generally very
> cautious with the use of the cloudbuster, said that he thought
> there might now be no danger in trying to have some rain.
> There were no clouds to be seen. He set up a cloudbuster,
> directed it toward a certain region of the sky, and took the
> caps off the tubes. He left it standing thus for about an hour
> and a half, then closed the tubes and took down the cloud-
> buster, and said that if he had conjectured correctly we might
> expect to have rain in some eight or nine hours. This was
> about 12:30 P.M. In the evening, toward 9:00, it started to rain,
> a mild drizzle that kept on until about 5:00 next morning. I
> tried to make some inquiries as to the extent of this local
> rain, and as far as I could find out, it had a diameter of some
> twenty to thirty miles. [Raknes, p. 75]

One of the few reputable scientists who tried to replicate
Reich's work on DOR and cloudbusting in a systematic way
is Dr. Charles R. Kelley, of Santa Monica. His findings were
published in 1961 in a special research report entitled *A New
Method of Weather Control*, which presented experimental
verification of Reich's weather-control technique. Dr. Kelley,

a student of Reich's, had been a weather forecaster during World War II. (He was later to become an applied experimental psychologist and director of the research laboratories of a major United States research firm.) In 1955 Reich printed a major theoretical article of Kelley's on weather; in it Kelley expands on many of Reich's concepts and ties them into existting scientific knowledge in a way Reich never attempted.* At the same time, he takes exception to certain of Reich's concepts. His discussion is too technical to summarize here, but will be of interest to other scientists interested in Reich's weather-control theories.

Another development emerged from the apparent DOR effects around Rangeley in 1952. This was the creation of a device to withdraw DOR from the human organism. Reich reasoned that if radioactivity could change orgone into DOR, possibly the inhibitory effects of muscular tensions and spasms would produce similar effects—that is, that the inhibition of the normal orgone flow through the body might turn it "sour" and dangerous to the organism. (Basically, DOR accumulation is the energetic condition which results from dammed-up orgone circulation through the body in interaction with its regular metabolism; one result is excessive retention of $CO_2$ associated with inadequate expiration; in short, the organism fails to take in enough orgone or discharge it properly.) He therefore produced a device to try to withdraw this hypothesized DOR from key tension spots in the body and called it a DOR-buster. It works on the principle of drawing off DOR (through water's hunger for orgone) using a device consisting of a metal funnel attached to BX tubing leading to nearby flowing water. One recent (homemade) version has the individual sitting or lying under an orgone blanket;† a metal funnel, perhaps a foot or two long, connected with BX cables is placed near supposed armored areas of the body. Another version used by some American orgonomists looks like this:

---

* Kelley, C. R., "Orgone Energy and Weather," *CORE*, Vol. 7 (1955), pp. 54–67. Included also as an Appendix in *A New Method of Weather Control*.
† The construction and use of orgone blankets are described in full in Chapter Sixteen.

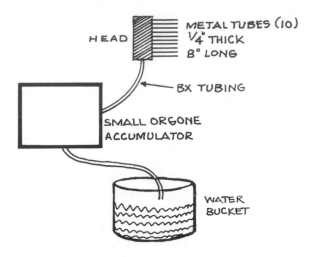

(The therapist passes the head of the device back and forth over the patient's body for about fifteen minutes in a treatment.)

When successful, this device provokes outbreaks of strong emotions supposedly locked in by the bodily armoring and thus facilitates therapy. While the device has been used by some doctors as an adjunct to therapy, it is uncertain how many orthodox Reichians either believe in or utilize it. Raknes says its use is still very much in the experimental stage and that "it should be used with the greatest caution and preferably only by an experienced [medical] orgonomist" (Raknes, p. 35).

As a consequence of both his rain making and his theorizing about DOR concentrations around Rangeley, in 1954 Reich became intrigued by the problem of deserts, desert vegetation and desert revitalization. He came to the conclusion that not only lack of moisture, but undue concentrations of DOR, were probably jointly the cause of some, if not all, of the world's deserts. His observations around Maine seemed to show that if DOR is prevalent in an area for a substantial period of time, trees and shrubs wither from the top downward and from the bark inward, and many die. The resulting lack of healthy vegetation in an area leads eventually to a desert condition, with soil erosion and the emergence of plants, like cactus, that survive only in desert conditions. Part of this process was confirmed by Dr. Kelley, and his paper on weather control includes photographs of trees going through this dying process. Reich also theorized that cactus and other typical desert plants are organic-plant counterparts to biopathic personalities—prickly and heavily armored—and that they were the last survivors in DOR congested areas undergoing this process of desert formation. The year before he was imprisoned he left Rangeley to test this theory and experimented with rain making in a part of the Arizona desert. According to his reports, carried in his journal, *CORE* (Vol. VII, March, 1955), this experiment was largely a success—that is, rain fell and the desert showed fresh vegetation. Lack of funds and time (Reich was then very heavily involved in the legal battle with the Food and Drug Administration) cut short the experiment. Although he advised the United States government of this endeavor, they paid no heed.

Similar disregard has attended the efforts of highly qualified professional men who have tried to secure funds to replicate various aspects of the orgone's applications. Dr. Bernard Grad, biologist at the William Allan Memorial Hospital in Montreal—whose experiments on healing energy will be described in Chapter Twelve—has run into a stone wall in efforts to secure funds to replicate Reich's work. In the United States, Dr. Charles Kelley, who is sufficiently well known to have a biographical sketch in *American Men of Science*, was turned down by fifteen different foundations when he sought funds for repeating his weather-control experiments. The tide may now be turning; reports in the English Reichian journal, *Energy and Character*, note that Dr. Hoppe, of Tel Aviv, is arousing some interest in weather control in the Israeli government and that Professor Giorgio Chiurco, noted Rome teacher of surgical pathology, has "succeeded in interesting President Makarios . . . owing to the long-standing drought problem in Cyprus" (*Energy and Character*, Vol. 1, No. 1, p. 49).

At first glance Reich's theories on DOR and weather control seem far-fetched, a tribute more to his fertile imagination than to his scientific ability. Yet if static electricity is accepted as basically the same as the orgone, there may be some basis for taking these theories seriously; for we know that clouds build up massive amounts of static electricity, periodically discharged in lightning storms. Also, the various devices or tools that Reich invented, up to and including the DOR-buster, while on the surface peculiar or perhaps ridiculous, should not be dismissed too lightly. Most new breakthroughs in science involve or focus upon a new machine or tool for investigating nature, and in the beginning these are clumsy or bizarre and often appear to outsiders as highly improbable.

In his highly erudite book *Scientific Knowledge and Its Social Problems*, British Professor (of the History of Science) Jerome R. Ravetz points out: "A problem under investigation grows in interaction with its materials, and when it is completed it is necessarily roughhewn . . . genuinely new experimental work frequently involves using tools at, or beyond, their limits of reliability, especially when they are used by untrained or unsympathetic workers." In a revealing footnote, he notes the difficulties Galileo had in getting acceptance for the telescope: "The existing 'spyglasses' that stimulated his invention were incapable of improving on the naked eye for astronomical purposes; and Galileo needed to control the quality of the fine glass produced at Venice and also devise his own system of lens grinding. Even then, most of the lenses he produced were inferior; and for the first crucial years, he simply did not have enough copies of a really good telescope to satisfy all the demands for demonstrations. The distinguished astronomer Magini brought a committee of professors to look through Galileo's telescope, and it was not difficult for them to see nothing of what he claimed was there." (Ravetz, *Scientific Knowledge, and Its Social Problems*, p. 266).

Ravetz emphasizes as well the amount of daring and faith it takes to make radical new discoveries. "The conquering of new pitfalls, the forging of new tools, and the establishment of new objects of inquiry, require great talent, daring and ruthlessness, and also a complete identification of the scientist with the result" (*ibid.*). In the following comment he throws some light on the necessary roughhewnness of theories at their early stages, such as that of Reich's: "If every anomaly in experience and every ambiguity in concept were completely ironed out before the work was presented to the public, nothing new would ever appear" (*ibid.*).

We know that history shows that all strange new discoveries run into various institutional blocks to acceptance, especially in the touchy fields of medicine and religion. Even simple technological novelties like the EEG machine took fifteen or twenty years to win acceptance in medicine. Ravetz adds a penetrating observation here: "If the problem has a bad history of being the province of cranks and speculators (e.g., mesmerism and healers), then whoever tries to rehabilitate it will have to struggle against a natural and justified prejudice. Worse, if the problem has a history of involvement in a political or professional struggle, then any advocate of it runs the danger of being dragged down with it" (*ibid.*, p. 267). Apart from Reich's own difficulties with the professionals in the international psychoanalytical association and with the psychiatrists in America, the problem he was studying did have a history of professional indifference. We must only conclude that regardless of the results of early experiments on weather control, or even simpler aspects of orgone application, getting a respectable hearing for such unusual conceptions will necessarily be, for some years to come, an uphill struggle, particularly among those professional groups which have most to gain by denigrating the theory and its various practical applications.

*Chapter Twelve*

RECENT EXPERIMENTS IN ORGONE
RESEARCH

After a long legal battle, Reich was sent to prison in 1956 for contempt of court, and he died there after serving nine months of a two-year term. His trial and imprisonment dealt a severe blow to the medical orgonomy group and the burning of most of his books by the government killed the dissemination and testing of his ideas—temporarily. In 1960 a book called *Selected Writings of Wilhelm Reich* came out in the United States, to be followed later by reprints of other books. Then certain individuals in the United States and small groups in Italy, Israel and Britain began in the sixties to revive interest in his writings. Some carried out scientific research. By the end of 1970, the tide seemed to have turned. In the United States there was Ellsworth Baker's *Man in the Trap*, on character disorders and Reichian therapy, Mrs. Ilse Ollendorff Reich's biography, a small book by Dr. Ola Raknes, Orson Bean's *Me and the Orgone*, and several others, besides publications in Britain, Denmark and Italy. Many reprintings were also now available. Serious journals like *The New Left Review* were looking at Reich's ideas and one, *The New Society*, carried an article "The Return of Reich." In addition, Britain's internationally respected psychiatrist, R. D. Laing, had given a favorable review* of one of the reprints and it

---

* "Liberation by Orgasm," *New Society*, March 28, 1968.

was rumored that he was intently studying all of Reich's writings.

Meanwhile, other researchers in the sixties turned out serious articles, papers or books which seemed to confirm one or more separate aspects of the orgone theory. Most of these were independent persons in no way connected with medical orgonomy. A few of these findings flowed from the burgeoning interest in parapsychology. Although some of the data are repetitive, since they help to confirm elements of the orgone theory, this chapter will report most of the relevant serious scientific activity from Western nations.

One of the most convincing series of scientifically impeccable experiments pointing to the existence of healing powers in the hands and the role of positive feelings in healing was directed by Dr. Bernard Grad, of McGill University, whose everyday work involves hormonal studies of the aging process. Some of his experiments on healing involved treating wounded mice by laying on of hands, the major experiment being conducted jointly with Dr. R. J. Cadoret and G. I. Paul in the Department of Physiology of the University of Manitoba, in Winnipeg, after preliminary study by Dr. Grad in his own time. Later on, Dr. Grad extended his studies to include plant and yeast cells. Eight articles describing the related experiments appeared over a period of a decade in such periodicals as *Corrective Psychiatry & the Journal of Social Therapy*, the *Journal of the American Society for Psychical Research*, the *International Journal of Parapsychology*, *Pastoral Psychology* and the *Journal of Pastoral Counselling*. Control groups were employed in such a way as to rule out all possible extraneous factors such as heat from the hands and suggestion.

An article in *Pastoral Psychology*, of September, 1970, summarized these various experiments. The first experiments, begun in 1957, used a self-declared healer who claimed to have successfully treated various diseases for ten years in his home town in Europe. Seventy mice, divided into two control groups and one treated group, were utilized. The treated group received the laying on of hands. One of the control groups received a daily exposure to the same temperature as was given off by the healer's hands, while the other received no "treatment" at all. The healer's treatments consisted of his holding a cage of eight to ten mice, one hand below and the other on a wire mesh over the mice for fifteen minutes, morning and evening, five days a week for forty days. (The mice had all previously been conditioned to remain relaxed in these small cages.) All the mice were exposed to goiter growth through feeding them a special chemical and a diet deficient in iodine. The goiters so induced were weighed at the end of the forty days.

> The results showed that the thyroids of the two control groups increased in weight significantly faster . . . than did the mice receiving the laying on of hands. Inasmuch as the thyroids of the control mice receiving the heat treatment did not grow significantly faster than did those of the non-heated controls, the heat produced by the warmth of the hands during the laying-on treatment could not be responsible for the significant inhibition in the rate of goiter development in the mice treated by laying on of hands. [Grad, in *Pastoral Psychology*, p. 20]

In a second experiment, designed to rule out the heat factor conclusively, wool and cotton cuttings were held first by the healer and then put into the cages of the experimental group while the controls received nontreated cuttings of the same size and material. "The cuttings were dropped into the cage and an hour later the mice were always found sitting on the cuttings . . . The results were the same as in the first experiment. That is, the thyroid glands of mice treated with cuttings held in the hands of the healer developed more slowly than did the control mice" (*ibid.*, p. 21). This finding supports the notion that the healing energy (orgone) can be transmitted to and retained in organic materials.

In another set of experiments small wounds were made on the backs of mice. In the first of these,

> treatment consisted of Mr. E.'s [the healer] holding caged mice between his hands for 20 minutes twice daily. The control group was allowed to remain in a similar cage without handling. Highly significant differences were found between mean wound areas of the treated and control animals on both the 11th and 14th days following wounding. The mean wound

areas of animals treated by Mr. E. were smaller than the control means. [Grad, Cadoret and Paul, p. 6]

A double blind experiment was added in which a special control group, who professed no healing abilities, also held the mice. Wound sizes were measured on the fifteenth, sixteenth, eighteenth and twentieth days following their infliction. A small but statistically significant difference in wound sizes occurred in the healer's experimental group as compared with both control groups. The factor of "gentling" was ruled out as all the animals were handled and "gentled" by their keepers before but not during the experiment.

In later experiments the healer held a bottle of saline solution (1% sodium chloride) used to water barley seeds. It was hypothesized that his energy would impregnate the saline solution and cause the seedlings thus watered to exceed the growth of a control group. Barley seeds were placed in twelve experimental and control peat pots alike. The watering followed double blind procedures. On the fourteenth day the height of the seedlings was measured.

> The results of the four experiments show that the mean height and yield of the seedlings watered with saline held between the healer's hands were consistently higher than that of the controls, and in three of the four experiments significantly so. . . . As a whole the findings indicate that something that acts favourably on the growth of barley seedlings passed through glass from [the healer] to the saline solution. [Grad, in *International Journal of Parapsychology*, Vol. 6, No. 4 (1964), pp. 485–86]

In a word, since this energy passed through glass and was accumulated in water for days or even weeks it was not any known form of energy.

Another set of fascinating experiments examined the effect of psychiatric states on plant growth, using a lab assistant as "healer" and two persons suffering from serious depressions. It was hypothesized

> that there was a direct relationship between the mood of the person doing the [laying on of hands] . . . on the solutions and the subsequent growth of plants watered by these solu-

tions. Thus, it was hypothesized that a solution held for thirty minutes in the hands of an individual in a confident mood would permit plants watered by this solution to grow at a faster rate than plants watered by identical solutions but held for the same time by persons with a depressive illness or not held by anyone (the control group). The experiment also tested whether solutions held by the depressed persons would inhibit plant growth relative to the control group. [Grad, in *Pastoral Psychology*, p. 22]

The results showed that the healer who held the saline water solution and one of the depressed persons who seemed to enjoy the holding of the bottle, nursing it in her lap, got better results, statistically, than the control. At the same time, the man with psychotic depression "produced" growth below that of the control group.

In another wound-healing experiment, first-year medical students whose attitude toward healing powers was understandably *skeptical,* and who laid hands on mice, produced a rate of healing consistently below that of those mice receiving *no* laying on of hands. In short, emotions of confidence or enthusiasm, depression or skepticism seemed to pass on to the saline solution different vibrations with differential healing consequences. Such a finding supports Ravitz' discoveries of how mood affects the field force, and the views of aura "readers" on auric emanations and feelings. Their relevance to orgone analysis is patent.

In 1967 a highly qualified psychiatrist, Dr. Shafica Karagulla published a book, *Breakthrough to Creativity,* which described her discoveries about healing powers, auras and magnetic fields around crystals. Dr. Karagulla, a neuropsychiatrist, "spent twelve years evaluating and studying mental patients, over five years of this time at the University of Edinburgh under the well-known British psychiatrist Professor Sir David K. Henderson . . ." (Karagulla, p. 24). Later, in Canada she spent three and a half years as Dr. Wilder Penfield's associate, evaluating patients with temporal lobe epilepsy and other serious mental disorders. Reading the life of Edgar Cayce "made a hole in the dike of [her] scientific mind" (*ibid.,* p. 26). She finally decided to investigate persons with what she called Higher Sense Perception (HSP), and read and traveled widely on this project for several years. Her book describes in a very winning way people she met who displayed HSP. One was a Dr. Dan,* who after convincingly diagnosing one of her physical complaints by aura reading described some of his activities:

> He admitted that he was able to diagnose a patient by observing the field of force which he could "see" around the patient. He was careful never to let the patient know this, and he always checked, using the normal examination procedures and laboratory tests. He had a reputation for being a most remarkable diagnostician. In addition he possessed some type of magnetic healing. This produced amazing results with children who had had infantile paralysis. . . .
>
> In the physical body he could see where nerve currents were blocked or not moving in a healthy fashion. Under such conditions he often applied magnetic healing energy and could observe the effect on the nerve currents.
>
> With regard to the endocrine system Dr. Dan could see moving vortices of energy associated with each gland. He looked for certain types of disturbances in function or for pathological conditions, depending upon the type of disturbance in the vortex of energy. [*Ibid.*, pp. 61–62]

Another physician with an outstanding reputation as a diagnostician practices in New York. He has found that a sensitivity in his hands indicated trouble spots in the body. After accidentally discovering with his children how he could relieve pain by putting his hands on the area affected, it gradually became

> a part of his routine to unobtrusively use this magnetic healing ability to relieve patients when it was necessary. When I inquired about any other ability he told me that on occasion he could see an energy field interpenetrating and surrounding the human body. He had not tried to develop this ability nor had he tried to correlate the field with any physical condition. [*Ibid.*, p. 66]

---

* The names used were all pseudonyms.

Dr. Karagulla tells of a practicing psychiatrist in New York who

> for years . . . had been aware of an energy field around people, which puzzled and perplexed him. He saw the field more clearly and somewhat extended around the fingertips. He could also see energy flowing in and out of the body and he saw central points where streams of energy crossed in the area of the spine. . . . He was relieved to know there were many people who could see this energy field and overjoyed to find another psychiatrist with whom he could talk. [*Ibid.*, pp. 71–72]

Numerous other medical specialists, including a surgeon, described similar gifts in her book.

On a grant from the Pratt Foundation, Dr. Karagulla checked into the gifts of certain outstanding clairvoyants. One, called Diane, is the president of a substantial corporation. She has frequently worked with doctors on medical diagnoses and they have verified that what she sees in the aura is actually there. For instance, Diane once met the famous writer Dorothy Thompson and in a few minutes diagnosed a serious blockage in her colon. Dorothy Thompson phoned her doctor for the results of recent X rays and found that they confirmed the diagnosis. In a few days she underwent an operation for an obstruction in the colon in the area pinpointed by Diane. From the following paragraphs it will be seen that Diane's clairvoyance reveals energy configurations like the Hindu chakras and bodily divisions like Reich's segmental analysis.

> Within this energy body . . . she observes eight major vortices of force and many smaller vortices. As she describes it, energy moves in and out of these vortices, which look like spiral cones. Seven of these major vortices are directly related to the different glands of the body. She describes them as also being related to any pathology in the physical body in their general area. . . . Each major vortex as she describes it more minutely is made up of a number of lesser spiral cones of energy, and each major vortex differs in the number of these spiral cones.

Five of these macro-vortices are located in a line along the spine. There is one at the base of the spine, one approximately midway between the pubic bone and the navel, one at the navel, one at the level of the mid-sternum near the heart area and one near the larynx or Adam's apple. There is another macro-vortex on the left side of the body in the area of the spleen and pancreas . . . There are two other macro-vortices, one approximately where the eyebrows meet and one at the top of the head. There is a ninth smaller vortex at the back of the head in the vicinity of the medulla oblongata. [*Ibid.*, pp. 124–25]

Insofar as one can accept the observations of an authentic and medically tested psychic, Diane's statements testify that Reich was right in seeing the body as an energy system, divided into segments, in which pathologies are related to specific segments. In addition, her observations—and they are supported, generally, by dozens of other reputable clairvoyants —suggest energy potentials at the chakras that lend credence to yoga claims and the kundalini theory.

In a chapter entitled "Three Energy Fields Around Human Beings," Dr. Karagulla provides further data from authenticated sensitives corroborating aspects of the orgone's activity. Her sensitives described three interpenetrating fields: the vital body, extending one to two inches out; the emotional field, extending one to one and a half feet beyond the body; and the mental field, another half foot or so farther out. Certain activities, like meeting a well-beloved person, brighten up and intensify all three fields. Moreover, she writes, "Some people achieve an access of energy observable in the brightening of their force fields from being out on the ocean or in a forest or from art, or music or creative work" (Karagulla, p. 161). This fits in with Reich's stress on the expansive effect on the orgone field of swimming in the ocean, walking through a forest or giving oneself to creative projects.

Having read von Reichenbach on the odic force, Dr. Karagulla also tested several of her clairvoyants on crystals and magnets. Diane confirmed that she saw force fields around crystals, metals, et cetera, and consented to some tests. She correctly pointed out the poles of unmarked magnets and vividly described varying and distinctive force fields around crystals they looked at in the American Museum of Natural History, in New York. By and large, these tests supported some of von Reichenbach's basic claims.

While government persecution suppressed orgone experimentation in the United States for years, the sixties saw the beginning of supportive experiments in other countries. In Argentina, one doctor has written about his extensive experiments. In Denmark, a leading psychiatrist, Dr. Tage Philipson, published *Love-Life, Natural and Unnatural*, which gave prominence to Reich's orgone research. In November, 1968, Walter Hoppe, who wrote about his use of twenty-layer accumulators in the *Orgone Energy Bulletin*, was invited to give a lecture at the International Cancer Congress in Cassana Junior, in Italy. He presented two papers on the development of cancer and the treatment of a malignant skin disease by the accumulator. In November, 1969, he was nominated a member of an important Italian medical society in recognition of his work in cancer therapy.

Reich's work was first introduced into Italy around 1960 by Professor Luigi de Marchi, a writer and sexologist. British neo-Reichian David Boadella says:

De Marchi holds a position in Italy today somewhat comparable to that occupied by Reich's colleague, J. H. Leunbach, in Denmark in the nineteen thirties. Leunbach then was a prominent worker in the sexual hygiene movement in Denmark. Luigi de Marchi was responsible for the formation of the Associazione Italiana per l'educazione Demografica in 1952. This association, of which he is still the national secretary, remains the only family planning organisation in Italy. In the teeth of Catholic opposition to the spreading of birth control information and advice, he and his wife have taken part in a dedicated struggle over the past sixteen years to establish the basic right of every mother to determine how many children she wants to have. This work has been at considerable personal cost, as de Marchi has been brought to trial on three occasions for these activities, and on at least one occasion substantially fined. De Marchi . . . is interested profoundly in the socio-cultural scene . . . and in particular in exposing the fallacies in the reasoning of the Marxist politicians and conservative-minded psychiatrists. [Boadella, p. 47]

After discovering Reich's work in 1959, de Marchi began publicizing it in Italy:

> In 1961 he arranged for Reich's *Selected Writings* to be published in Italy. Translated and introduced by de Marchi, this book was the first by Reich to appear in Italy, under the title *Teoria dell' orgasmo e altri scritti* (Milan: Editore Lerici, 1961). The appearance of this book aroused lively controversy in the Italian press, and a miniature newspaper campaign was mounted in two important Italian journals, *Il Tempo* and *Paese Sera*. [*Ibid.*]

Boadella points out that de Marchi refuses to parrot Reich's ideas, but takes a broad sociological approach to them. Around him grew up "a loosely formed group of seriously interested professional people" (*ibid.*, p. 48). One of these was psychiatrist Dr. Bruno Bizzi, vice-director of the Psychiatric Hospital, of Imola, near Bologna. From the beginning he was interested in the "application of orgone energy to some of the psychosomatic complaints shown by his patients at Imola. Three years ago Dr. Bizzi secured the permission of the director of the hospital, Professor L. Telatin, to introduce some orgone accumulators into his work there, and he has since then claimed beneficial results in cases of rheumatism, neuralgia, and particularly cancer. The results with cancer were so promising that they became known to Professor Giorgio Chiurco, former head of the department of surgical pathology in the University of Rome, and the director of the Centro Sociale Studio Precancerosi e Condizioni Premorbose (CESPRE) in Rome. Professor Chiurco is an internationally known researcher on cancer prophylaxis, and his sympathetic interest in Dr. Bizzi's work meant that within the past few years a large number of doctors have become interested.

"At Rome in 1968 (4–7th October) the second international seminar on the prophylaxis and prevention of cancer was held. This was a truly international gathering, with representatives from more than 60 countries. At this conference Dr. Bizzi presented a report on 'energia orgonica—forza vitale (Galvani) e stati morbosi' " (ibid., pp. 48–49).

In this paper Dr. Bizzi said, "We have . . . been able to confirm the famous increase in temperature inside an experimental orgone energy accumulator" (Bizzi, in Energy and Character, January, 1970, p. 57). Further, with respect to the accumulator he adds:

> We have obtained good results when dealing with pain of rheumatic origin, with some kinds of neuralgia, constipation and insomnia. In the psychiatric field, it has proved useful in treating anxiety and neurasthenia . . . the radiation has in fact a precise vagotonia effect which relaxes the muscles and treats the sympatheticotonia which is an essential basic component of various morbid states on the symptomatological level . . . It is important in the majority of morbid cases, that treatment should be prolonged and very regular; we have never witnessed miracles and we have recorded failures in chronic cases which had already resisted other treatment. [Ibid., p. 58]

Specifically with respect to cancer, he accepts a bio-energetic analysis; in particular, he views frigidity as a precancerogenesis phenomenon along with the shrinking process. From research on many (thousands) of women, suffering from neurotic and somatic disturbances, he notes that in this group, 92–94 percent were frigid. He specifies that "in this group of people, alongside anxiety and contraction, we find thick fibroids and growths in the womb, lumbago, rigidity of the sacral lumbar tract of the spine and pathological curvature of the spine . . . finally, an ever-present disturbance is always the inhibition of expiration with a chronic inspiratory attitude and diaphragm block" (ibid., p. 60).

Along with Reich, he identifies the medical basis of these conditions as anxiety. And in his address he emphasized that anxiety "is accompanied by bio-energetic contraction, a concentration of fluid in the center of the organism, sympatheticotonia with hypertonia of the cardiovascular apparatus, spasms, diverse contractions in this or that organ of varying intensity and localization. Breathing is always reduced, with inhibition in particular of expiration" (ibid., p. 61).

At the same conference Professor Chiurco presented a paper subtitled "The Functional and Organic Biopathy of Man." Boadella tells us that at this conference "Professor H. J. F. Baltrusch, of the International Psychosomatic and Cross-cultural Leukemia Project, Oldenburg, Germany, and also of the Krebsforschungsinstitut, at Aarhus, Denmark, expressed great interest in Reich's cancer research to the extent of imploring Dr. Bizzi to send him photostats of 'The cancer biopathy' " (Boadella, p. 49). In 1970, Dr. Bizzi was invited to give ten lectures at the University of Rome on orgone energy and cancer prophylaxis. And in May of the same year, he and Professor Chiurco presented further papers on the bioenergetic approach to cancer at the World Cancer Congress in Houston, Texas.

Some new research in weather control and DOR grounding experiments in the United States throws a little more light on this more debatable aspect of the orgone theory. In the sixties, three Americans—namely, Dr. Charles Kelley, Dr. Richard A. Blasband and Trevor Constable—spent considerable time and money in checking on these experiments. All these men, who are well regarded in their fields, claimed to have confirmed Reich's results.

As indicated in the previous chapter, Dr. Kelley first began investigating Reich's weather-control theories in the early fifties. He was at first completely skeptical, having been engaged in forecasting weather on the basis of orthodox meteorology while serving in the Air Force Weather Service in World World II.* However, on a visit to Rangeley, Maine, he saw Reich aim the cloudbuster at a cloudbank, and the clouds dissipated in fifteen minutes. Subsequently he made his own apparatus and made many trials over a three-year period. He writes:

> Cumulus clouds of fair weather (cumulus *humilis*) could usually be wiped entirely from the sky in 10 or 20 minutes. Cumulus *congestus*, a cumulus cloud of greater vertical development, took longer but responded unmistakably, frequently disappearing entirely within 20 to 30 minutes, while control clouds continued to grow. The many repetitions made

---

* He was an N.C.O. in charge of two weather stations during the war.

the results unequivocal. [Kelley, *New Method of Weather Control*, pp. 3–4]

In his booklet on weather control Kelley includes a picture of the antenna portion of a Reich weather-control apparatus. He clarifies the theory behind cloud-busting as follows: "Since water vapor and droplets are attracted by orgone energy, the cloud is built and held together" (*ibid.*, p. 7) by its orgone-energy concentration—that is, it is an orgone energy system. By modifying its energy system, one is able to change or dissipate the cloud. To increase the size of the cloud, one must draw energy from around or beside it, thus increasing the energy differential between cloud and surrounding air and theoretically producing an increased flow of energy into it. Kelley admits it is much easier to destroy a cloud than to build it (*ibid.*).

Using time-lapse photographs, Kelley shows how his cloud-buster dissolved a series of growing cumulus clouds. The key scientific issue, especially in dealing with cumulus clouds, is whether their dissolution is not simply a combination of chance and wind action. Kelley argues that he utilized control clouds in his demonstrations, but few of the seven sets of photographs included in his report picture these controls adequately. While one cannot doubt his sincerity or scientific integrity, one has to be skeptical of this demonstration until the experiment is replicated and more convincing photographs of what happens to the nearby control clouds are produced.

After some years of skepticism, in 1959 Kelley began periodic DOR removal, when living in rural Westport, Connecticut. Previous experience with the apparatus had apparently resulted in unusual headaches and pressures in the head and so in this rural setting, Kelley used a garage, away from the house, to store his equipment, including orgone accumulators. Nevertheless, he reports:

> Gradually the atmosphere in the house became oppressive and unpleasant. Our health, including that of the baby, suffered severely. Our symptoms corresponded to what we had experienced previously when we had been forced to move, and to what Reich described as the consequence of dor infestation . . . The oppressiveness of the house continued, however, and

got worse, when winter forced us to close doors and windows. The oppressiveness seemed to be in some way associated with our heating system . . . In an attempt to improve the situation, I set up a small Reich weather-control apparatus in a corner of the basement, directing the antenna toward the furnace, and grounding the apparatus in a large jar of water, which I changed every few days. We were never sure whether this improved the atmosphere of the house, which remained extremely oppressive, but it did seem to result in much worse atmosphere in the basement. The apparatus was left in place for three months. Toward the end of this period (winter 1959–60), I noticed a remarkable change taking place in the stones in the foundation wall directly behind the antenna. The rocks were turning black. Elsewhere in the cellar the rocks were normal in color, mostly light brown or gray. The rocks affected were close to the rear of the dor-removal equipment and in the direction toward which the antenna was drawing.

It is just possible that the blackening was caused by some water leakage.

Reich has described the blackening of rocks as a specific consequence of severe dor infestation. When, according to his report, heavy dor infestation rendered his laboratory and observatory virtually unlivable (in the Oranur Experiment), the rocks in the walls and fireplace of infested buildings began turning black. The nature of the blackening process and characteristics of the blackening rocks which Reich described matched in every detail that which occurred in my own basement. The blackening is not the result of carbon deposit, which it resembles only superficially, but of a change in the surface structure of the rock itself. The blackened surface tends to develop a granular structure. Rocks first being affected frequently show tiny round black pockmarks before the condition spreads. It looks a little like a black mold at this stage, but is not soft . . . The black surfaces emit a radiation which feels warm and tingling to the skin. [*Ibid.*, pp. 19–20]

One wonders why this mold was not given close microscopic inspection.

Kelley now turns to another consequence of DOR:

Prior to the blackening of the rocks in the basement wall, the top of one of the trees in my yard, near the Reich weather-control apparatus, lost its leaves and appeared to be dying. I gave this no particular thought at the time, not knowing that Reich had also written of this as a specific effect of dor . . . The reappearing spring leaves showed that most of the trees close to the weather control apparatus were dying at their tops. The bark turned black and disintegrated or peeled, and branches died. Trees across the pond, or at some distance from the weather-control apparatus, were not affected, nor were any other trees of the thousands in our neighborhood. Yet nine trees of different kinds, situated within 60 feet of the experimental apparatus, were dying at the top. As the summer continued, the dead areas of these trees increased. Dead branches dropped. One tree had to be cut because it became a danger. [*Ibid.*, pp. 19–20]

Accompanying this article were convincing photographs of these dying trees.

Later on, Kelley read Reich's last book, published in 1957, and found there a description that tallied exactly with what he observed.

"DOR penetrates the trees slowly from the top downward, and from the bark inward. . . . We see at first the bark getting blackish; then the bark disintegrates, and disappears. The process never sets in from the roots upward; it is thus not due to 'bugs.' The disappearance of the bark regularly begins at the tree tops, working its way downward toward the roots. Also, the blackening and ensuing disintegration of the bark begins on the upper sides of the branches; this points . . . to the atmosphere as the source of the noxious agent." . . . The dead branches [that Kelley discovered] proved to be exactly as [Reich] described; their upper sides had turned black and had begun to disintegrate, while the bottom sides retained their normal color. [*Ibid.*, pp. 22, 24]

Kelley's explanation of this apparent noxious infestation is that the antenna attracted and grounded around the trees a certain amount of DOR. (He does not mention the accumulators nearby, which could have assisted in this concentration.)

He reasons that once a considerable orgone level is built up in an area it draws in more orgone (on the principle of orgone flows to high-potential concentrations). Then, if this becomes stagnant or immobilized, it can change to DOR. Readers interested in a full explanation of the strange phenomenon can find it on pages 24 and 25 of his weather-control booklet. For the purposes of this book it suffices that his experiments suggest that a powerful energy was present, sufficient to change the exterior of some rocks in his basement.

Kelley's booklet also recounts a few successful rain-making experiments carried out between 1958 and 1966. Since Blasband extended his efforts over a greater length of time and apparently succeeded in affecting an acute drought situation it will be more useful to describe his results without slighting Kelley's experiment.*

Blasband, who had just moved to a country home in Pennsylvania in 1965, was impressed at that time with signs of a drought then apparently overtaking New England. Lacking rain for several weeks, soil and vegetation were parched, trees seemed to wilt, the sky was a "hazy gray-yellow-brown" and the "rare clouds were then dirty-white and had ragged, tenuous edges." Even people seemed affected, "fingers and toes were swollen, tempers short," and some had a "metallic taste in the mouth and a general feeling of malaise" (Blasband, in *Journal of Orgonomy*, May, 1970, p. 66). (One meteorologist, writing about the New England drought from 1962 to 1965, had described it as a "major climatic aberration" [J. Nameas, p. 94].) In due course, Blasband diagnosed the situation as due to the presence of DOR, and began on June 13, 1965, to attempt to dissipate it and to produce some rain.

The problem was to demonstrate that rain came from the operation of the cloudbuster and not from chance. Blasband argues that the only reliable scientific controls are the forecasts of the U.S. Weather Service, normally 80–85 percent accurate for a twenty-four-hour period and 70–75 percent accurate for a forty-eight-hour period. His criterion of successful rain making was, accordingly, the production of rain (if only a

---

* The one dramatic experiment occurred on August 21, 1960, and is described on pages 40–42 of his booklet.

trace) within forty-eight hours of the start of the operation, if
the Weather Service's forecast was 10 percent or less probability of rain for the next twenty-four hours and no precipitation expected for forty-eight hours. He writes:

> In 1965, we found that, out of 38 operations to engineer
> rain under the above criteria, we were successful 18 times.
> After allowing for a 30% chance of mistaken forecast, we
> still find that this is 1.6 times more rainfalls than one would
> have expected by chance alone. Distributed throughout the
> droughty part of the year, these 7 additional rainfalls brought
> considerable relief. [Blasband, p. 69]

Whether Blasband's operations achieved statistical significance
remains open to question.

In 1965, however, his success with rain making, admittedly, was "equivocal at best," and for several reasons.

> One of them was the discovery that, in addition to *drawing*,
> the cloudbuster also *excites* the atmosphere. Experimental
> draws from large thunderheads would result at times, not in a
> shrinking of the cloud, but in sudden growth, an explosion to
> gigantic proportions. In time-lapse films, the cloud looked like
> a giant amoeba being irritated and excited. Similarly, a draw
> for rain would, under certain conditions, result not in rain
> but in intensification of drought. At such times, the atmos-
> phere would appear particularly excited and charged, and all
> our instruments would verify this impression. [*Ibid.*, p. 70]

Two comments are obvious: either that the failures were
proof the cloudbuster is a fake or that it links the two energy
systems of thundercloud and earth and when the latter is
stronger, by orthodox weather theory, it *should* draw orgone
from the earth and get bigger.

Rain-making attempts in 1966, in the same area, got off to
a discouraging start. Blasband is nothing but honest, when he
writes, "In April of 1966 the drought had set in again. We
found it impossible, using our previous techniques, to affect it
very much, apparently because of the exciting function of the
cloudbuster" (*ibid.*, p. 71). However, failing to draw by pointing the bloudbuster to the west—following orthodox Reichian
theory—he reasoned that perhaps pointing it to the south

would draw energy and moisture up from the Gulf of Mexico. So the cloudbuster was left pointing in that direction for long periods (called long "draws"). Immediate results were gratifying. For instance the forecast for Sunday, April 17, was "clear skies through midweek . . . We began the draw on the 18th and continued it . . . with one day's interruption until the 24th" (*ibid.*). The sky quickly clouded over, and it rained on April 19, 20, 22, 23 and 24. By "mid-May, the Northeastern U.S.A. had more rain than in the entire months of May 1964 and May 1965 combined . . . It was the heaviest rainfall in New York State for the month of May since 1953. By mid-June the drought had substantially decreased . . . There were reports of the rich flow of moisture coming up from the Gulf of Mexico" (*ibid.*, p. 72).

Then, Blasband admits,

unfortunately, it was not long before this rich stream of moisture began to dry up. Later in June all further attempts to generate rain not only failed but seemed to intensify the dryness and elevate temperatures. The atmosphere felt "feverish" with intense excitation. There was no rain locally, and all clouds approaching the area rapidly dissipated. Subjective perceptions and objective measurements indicated a return of chronic expansion . . . DOR began to build. [*Ibid.*]

Early in July, reasoning that perhaps they were accelerating the drought, he stopped rain-making operations. Where many would have claimed such failures proved the cloudbuster was ineffective and given up the experiment, Blasband produced reasons to "explain" the failure. He made three observations, the first being that Reich had noted that orgone gets results quickly in initial accumulator experiments, followed by a later "dulling" or only slight effects; so with the atmosphere, initial draws get good results, but later ones have minimal effects. While I had not noticed this observation about the accumulator elsewhere in the literature, it does tally with my over-all impressions in using the orgone blanket—that is, initial dramatic results seldom got repeated, but were followed by minor, although evidential, effects. Could it be that the orgone works to restore some natural equilibrium in the orgonotic field (which may result at first in dramatic effects) but that once this is largely established, only minor fluctuations from it are effected by subsequent use of an orgone blanket, accumulator or cloudbuster? Doubtless this is at best a partial explanation, but it does make sense of certain puzzling findings. As for the apparent DOR effects that Blasband noted, perhaps these were due to excessive draws of orgone into an area without adequate capacity for discharge.

Another comment may be important—namely, that in these experiments the apparatus was "discharging" into a well and not into running water. Blasband asks: "Could it be that a poorly grounded cloudbuster excites and draws without permitting adequate discharge?" (ibid., p. 74). To improve the effect, he filled the well and let water run into it constantly. Then trying again in September, 1966, and drawing to the southwest, results were immediately gratifying. Rainfall for October was 5.1 inches, compared with a normal of 2.8, and continued heavily up until the spring. Of course, by this time the atmosphere would have had a chance to normalize itself. Blasband points out, in a letter, that "the forecast for the next 30 days was a continuation of drought, and the storm that broke the drought rapidly followed our weather operations."

In 1968, Blasband reported success in clearing out DOR which he thought may have been unusually high because of a solar eruption that July. After using the cloudbuster he says, "Unfailingly within 24–48 hours a strong fresh energy stream from the northwest would 'chase out' the DOR" (ibid., p. 76). He also observed, "Draws for rain *following* a clearing operation were quite successful. Five out of six succeeded that summer" (ibid.). Later, referring to the spring of 1969, when he had ceased "weather engineering," he mentions northwest breezes that would "spontaneously clear out the [DOR-infested] atmosphere" (ibid., p. 77). This, of course, raises the question of whether similar breezes in the previous year were not spontaneous rather than the result of his cloudbuster.

Recent rain making was attempted in the Los Angeles area by Trevor Constable, a radio officer with the United States merchant marine and a historian. Diagnosing the situation there in early April, 1971, as droughty owing to DOR-ish conditions, he began operations on April 3, to test "the efficacy, under Southern California conditions, of Wilhelm

Reich's technique of drawing against the galactic orgone stream to produce rain. Prior experience in this geographical region, with extremely simple, rack-type cloudbusters had demonstrated that frontal systems could be 'fertilized' by judicious orgone engineering, and the necessary instability of the atmosphere that is deemed the *sine qua non* of rainfall by most conventional meteorological theory, could be created" (Constable, p. 192).

In an article subtitled "Rainmaking, L. A. Basin Area," he supplies weather maps and reports from forecaster Bob Hale (NBC-TV) for April 12, 1971, to indicate that no significant precipitation was expected. Without going into Mr. Constable's explanation of his rain-making technique it is perhaps sufficient to say that he began northwesterly "draws" at 10 A.M. on April 13, and by 6 A.M. on the fourteenth, rain began to fall. Constable shows hour by hour how his "draw" proceeded and how weather forecasts gradually changed. In spite of official meteorological forecasts he points out that "the entire region was drenched. Los Angeles Civic Center received .5 inch, Culver City, .66 inch, Santa Monica .75 inch; and 1 to 1.5 inches fell in various parts of the parched San Fernando Valley" (*ibid.*, p. 197). Headlines in the Los Angeles *Herald-Examiner* announced: "April Rains Surprise L. A." While one rain-making success proves nothing, the documentation of this specific case in the *Journal of Orgonomy*, with its hourly sequential notations of cloudbusting operations and broadcasts of changing weather conditions, is impressive.

A careful examination of Blasband's and Constable's reports, however, leaves one with many doubts respecting both the rain-making and DOR-clearing ability of the cloudbusting apparatus. Effective controls are difficult to achieve and Blasband's account seems subjective at critical points. (Constable says nothing about his efforts from April 4 to 12.) Blasband's honesty in admitting failures is impressive, but his claims leave one looking for other explanations. On the other hand, his warnings of a long-term trend to drought in the United States Middle-eastern and Northeastern states and the long-term effects of contamination of the atmosphere are well taken. In a word, while one must agree that the pollution of the atmosphere by both nuclear explosions and industrial

contaminants is resulting in a more or less permanent unhealthy atmospheric situation (which may eventually produce
more deserts), Blasband's tendency to use terms like *oranur*
and *DOR* to describe this process seems naïve and oversim-
plified. There may be truth in the orgone and DOR approach
to weather and drought conditions, but the evidence to date is
terribly thin. At times, in fact, one is inclined to see these
discussions of DOR and Reich's latter-day description of
flying saucers as malevolent cosmic orgone-energy machines,
as largely projections of a harassed and somewhat paranoid
mental state. Or, it might be that they represent the typical
temptation of great innovators to push their original insights
into distant fields where their appropriateness is, at best,
minimal.

Perhaps the most striking confirmation of the value of the
orgone theory in the sixties has been the extensive growth of
those bioenergetics-type therapy groups in various countries
which take for granted Reich's concepts of character armoring and orgone flow through the body. These took various
forms in different nations and regions. One of the most
interesting, called "Therafields," is in Toronto, Canada. This
embraces over five hundred adults who combine a bioenergetics program with massage training and eclectic individual
and group therapy aimed at getting the unconscious mobilized. Many of the five hundred reside in large houses and
work together, seeking to create a therapeutic milieu. The
group, while only six years old, is growing rapidly and draws
frequently on Dr. Lowen or Dr. Pierrakos for ideological
fertilization.

California has spawned numerous therapy institutes, many
of which borrow considerably from bioenergetics. These
include the Western Institute for Bioenergetics Analysis, in
San Diego. One of its members, Stanley Keleman, D.Ch.,
describes something of the "grounding" concept in the
quarterly *Energy and Character*, September, 1970.

> To me, grounding means being anchored in our physical psy-
> chic growth processes; expanding, contracting . . . charging,
> discharging. Grounding means being rooted in and partaking
> of the essence of the human animal function . . . we share

our basic modes of life with the other animals and respond in
the same way they do in our involuntary muscular and nervous systems. [Keleman, p. 11]

He thus expands on an idea of Alexander Lowen which builds
on Reichian concepts of the armor, its layering, and the need
to relate to the psychic "core," etc.

Another interesting California institute, promoting direct
body therapy, is run by Dr. Malcolm Brown, of Berkeley.
Brown uses many of the basic bio-energetics exercises, but is
skeptical of their potentiality for releasing tender emotions
and building up those with weak egos. He begins his therapy
by putting "the open palm of one hand gently against the
abdomen of the patient between the sternum and the navel"
(M. Brown, p. 1). Depending on the reaction of the patient,
the therapist may keep his hand there for from two to five
minutes. The aim is to promote deeper, more rhythmic
breathing, relaxaticn of the body's metabolism and mobilization of the energy charge and blood flow. Later, the hands
may be placed elsewhere around the torso—there are five
positions—to induce greater freedom of energy flow and emotional discharge. The over-all aim is to encourage a from-
within-out release of armored feelings, rather than their evo-
cation by pressures of the Reichian squeeze-and-pinch type.
Besides believing in the orgone flow, Brown accepts some of
Lowen's and Reich's techniques within the context of his
tender body-touching approach. It would seem his approach
tries to marry a kind of spiritual laying on of hands with
some of Lowen's exercises. The gentle stroking might well
mobilize latent feelings of love, but perhaps the Reichian
pinch-and-squeeze exercises would still be needed to mobilize
deep layers of fear and rage.

In Santa Monica is located the Interscience Work Shop, the
California facility of the Interscience Research Institute of
Connecticut, the organization that sponsored Dr. Charles
Kelley's work in weather control and published the neo-
Reichian journal *The Creative Process* in the years 1960–65.
Dr. Kelley, whose primary training is in psychology, now
lives with his wife, Erica, in California, and they operate the
Interscience Work Shop. The Kelleys employ Reich's deep-

emotional-release techniques with groups as well as with individuals. They have given neo-Reichian workshops and seminars around the world, at centers such as Esalen and Kairos in California, the Institute for Bioenergetic Analysis, the Aureon Institute in New York, and Qualsitor Institute in London.

There are two unique features to the Kelleys' work. First, they insist that their work is educational rather than medical or therapeutic; and second, they base their program on Dr. Kelley's studies of the nature and origin of muscular armor. The armor is a product of the imperfect evolution of the capacity for purpose, Kelley has concluded. It is not an illness to be cured, but an important function that has developed imperfectly, and requires educational work to perfect. Kelley writes:

> The blocks to feeling that Reich calls "the armor" and Janov "the defenses" are a product of the capacity of man to control his feelings and behavior, and so to direct his life along a path he has chosen. One aspect of this is protection of the self from incapacitating emotions, a second the channeling of behavior towards goals. Both are important to survival. The key function they involve is that of voluntary, as opposed to spontaneous, control of attention. This control is a negative process, a learning to block attention and movement in directions one does not wish to go. It is accomplished by selective contraction of muscles. It is the basis of man's capacity, first, to function under stress, and second, to live his life long-range, to have direction, self-discipline, purpose.
>
> I have described the process in greater detail elsewhere. Note the opposition between the goals of "education in feeling" and those of "education in purpose." "Feeling" is in the here and now, "purpose" is oriented to the future. Purpose is in important respects antithetical to feeling, and it is from the evolution of purpose that blocks to feeling have arisen. The "defenses," the "armor" are an expression of man's developing but imperfect capacity to control and direct his attention, his energy, his emotion, his movement, his life.
>
> The complete human being requires direction as well as feeling in his life, self-discipline as well as spontaneity, toughness

as well as tenderness. Emotional-release techniques leading to growth in the capacity for feeling are a worthwhile goal for most individuals, because the evolution of purpose has developed imperfectly, and has severely curtailed modern man's capacity to feel. Man needs to free himself from excessive and inappropriate blocks to feeling, to learn more effective and less damaging techniques for keeping his life "on track" and going where he wants to go.

Yet to abandon *all* blocks to feeling would be to abandon all self-direction and control. To function effectively for a single normal day requires that man sometimes block his immediate impulses, his feelings, that he use voluntary as well as spontaneous attention. And voluntary attention is forced; it is the function underlying purpose, the defenses, the muscular armor. [C. R. Kelley, *Primal Scream and Genital Character*, p. 8]

What may be a breakthrough for Reichian therapy in Britain occurred in January, 1970, when psychoanalyst J. W. T. Redfearn, M.D.O.P.M., chairman of the medical section of the British Psychological Society, addressed his colleagues on bodily experience in psychotherapy. Dr. Redfearn's lecture dealt with the bodily implication of bioenergetics techniques and was aimed at relating these to Jungian concepts of the body image. Redfearn had attended seminars on vegetotherapy and had experienced a bioenergetics workshop led by Dr. Lowen. Commenting on this, he said:

These people (Lowen & associates) say that stony hard tense muscles correspond with deeply repressed affects. Such muscles need much work before they become soft and can relax, when the conflict is said to come to consciousness and abreaction and remembering take place. I have no reason to doubt this . . . In the first place, I was rather impressed by Dr. Lowen's ability to use his observations on body postures, slight deformities and tensions, together with psycho-dynamic interview methods to produce a very telling diagnosis on the character defenses of those present at his workshops.

After telling how he has enlarged the scope of his therapy to include direct body work, "using a direct massage of the most rigid areas of the body," Redfearn reported such changes

as "some lessening in . . . bodily frigidity and stiffness. Con-
comitantly there were changes in facial expression, in that the
face became more alive, and the eyes lost their dull implacable
look." He concluded part of his paper saying, "I am glad of
this opportunity of learning more by means of practical work
about the so-called body armour and its physiological effects.
I wonder if what we can learn from the followers of Reich
about body armour we can repay by our greater knowledge
of ego splitting and body mind splitting" (Redfearn, in *Character and Energy*, pp. 75–76).

In effect, Dr. Redfearn was recommending to his audience
that they give time and study to the Lowen techniques and to
the whole idea of body armoring and its effects on the emotional and physical state. Whether Dr. Redfearn's lead to the
British therapists will take root or not is uncertain; he himself said in the lecture: "We analysts are only a tiny bit less
timid and rigid than our parents, or the man on the street"
(*ibid.*, p. 76).

What is significant is that the fast-growing new interest
in the body and its role in health, sensitivity and creative
living is emphasizing the validity of Reichian and neo-
Reichian therapies. As more and more intelligent therapists
and doctors experiment with ways of bringing the body and
mind together and releasing long-standing feelings locked in
rigid musculature, more will be learned about the energy flow
through the organism and how correct Reich's formulation of
it was.

Perhaps the best way to round out this discussion of new
developments is to report briefly on a new American organization which is designed expressly to explore the life-energies
theme. This organization, Life Energies Research, Inc., held a
founding conference at Wainwright House, Rye, New York,
in November, 1970, on the theme "Exploring the Energy
Fields of Man." Attended by twenty hand-picked representatives of the sciences, parapsychology and the paranormally
gifted, "the main purpose of this interdisciplinary exchange
was to review all known energies connected with the aura and
healing" (Report on Conference, Exploring the Energy Fields
of Man, p. 1). One conclusion was that further knowledge of
the physical universe might lead to eventual explanations of

parapsychological and other unusual energies by physical laws.

It is suggestive that the physicists at the conference showed a considerable open-mindedness to parapsychological energies like psychokinesis, and to explanations of how psychics use the aura, et cetera. They were concerned to discover how psychics differ constitutionally from the rest of us. The report says, "It was found that there are many top-level physicists . . . who no longer dismiss serious inquiry into these phenomena as the activities of crackpots" (*ibid.*, p. 2).

The conclusion of the report hints at the possibility that this highly responsible organization* is moving in the direction of at least one basic orgone postulate. For we read,

> the first decision reached by LER [Life Energies Research] following the conference was to sponsor a research project in psychic healing which is now being carried out by a research psychologist. Attempting to find out what potential exists in "ordinary" people to "feel" what the healer claims to feel when engaged in the act of healing, training seminars are being held regularly. The theory behind this experimental research is based on the results of a thorough study, previously engaged in, of the similarities of insights which exist among "sensitives" (psychic healers), physicists and mystics. These studies revealed a similar "world-picture" of the basic unity of all things . . . Since this "oneness" implies interconnectedness and an exchange of energies, it would seem that we all live in a "sea of energies" . . . It can be said that early reports on this research indicate that there are such potentials in all of us. [*Ibid.*, p. 3]

In short, the Reichian belief in unusual energies in man, and their relationship to our rootage in an ocean of energies (orgone) will now be tested by a quite independent and scientifically respected group in New York.

---

* Its board of directors includes Andrija Puharich, M.D.; Robert Jeffries; Dr. Eng., past president of Data Control Systems; and John E. Laurence, a nuclear physicist, influential businessman and member of the Advisory Committee during the initial establishment of N.A.S.A. The director is the respected New York psychiatrist Dr. Robert Laidlaw.

PART

V

*Chapter Thirteen*

# Orthodox Science and Unusual Invisible Energies

One way to begin an assessment of the distinctiveness and reality of the orgone is to survey what present-day orthodox science says about invisible energies that affect organisms and see to what extent the orgone "fits in." Since the orgone is regarded as having healing properties on the one hand and being intimately related to the atmosphere and to energy components in our solar system on the other hand, it will be logical to examine accepted scientific thinking and research on these two aspects of invisible energies in turn.

All matter, unless at absolute-zero temperature, vibrates and radiates. Daniel S. Halacy, Jr., a well-known science writer, notes that "radiation (the process of emitting energy as electromagnetic waves in the form of waves or particles) seems to be the pulse of the universe" (Halacy, *Radiation, Magnetism and Living Things*, p. 25). This is because all matter possesses an electric charge. And when matter vibrates or oscillates, it produces electromagnetic waves. These vary in size from wavelengths of extreme smallness to those of vast length. At one end of the scale there are gamma rays, some as short as one ten-billionth of a centimeter in length. At the other end are others, some eighteen million miles long.

An electromagnetic wave consists of an electrical field with a magnetic field at right angles to it. The entire wave moves at right angles to both these fields. One recent theory describes this radiation as a "wavicle," a compromise between the specific-particle and wave theories. A well-known physicist, George Gamow, describes electromagnetic radiation as a jelly-like material or "perhaps a kind of cloud" (Halacy, p. 26). Since these radiations contain energy, they can affect living organisms. But it is not the amount of energy "sent" into an organism, but the amount absorbed by the cells that is effective (Ellinger and Lanz, p. 4). Simple examples of use of electromagnetic waves in therapy are deep-therapy lamps. These give a variety of wavelengths including the infrared waves. Infrared and diathermy machines now commonly used in hospitals for specific conditions utilize electromagnetic radiation in the short-wave spectrum—from a few centimeters to about fifty meters in length—and have accepted therapeutic value. In addition, it has been found that microwaves are apparently therapeutic with quite a range of conditions, from arthritis and spondylosis (Binstoc, p. 184) to cerebral sclerosis, rheumatism and endocrine disorders, as well as bacterial and viral inflammations (Schliephake, p. 23).

It has long been known that *magnetic* energies have effects on living plants and animals. It was from such experience that Mesmer got much of his inspiration, albeit it was unscholarly and extravagant. In recent years new interest has been evinced in this area by some orthodox scientists. Researchers in both Russia and the United States have demonstrated that a magnetic field increased the growth rate of tomato plants. "This phenomenon is called magnetotropism. The researchers placed magnets about small ripening tomato plants and showed that they . . . grow faster in a magnetic field. [Moreover they] ripened relatively faster when nearer to the south pole of a magnet than to its north pole" (Halacy, p. 60).

Dr. M. F. Barnothy brings together in his *Biological Effects of Magnetic Fields* a wealth of data on this subject. He emphasizes that the effects are closely related to the strength of the magnetic field along with the exposure time of the subjects. Some of the results he mentions are: arresting of growth of mice kept in a magnetic field of 5,900 Oe;* reduc-

---

* "Oersteds"—units of measurement of intensity of a magnetic field.

tion of white-blood-cell count in mice so exposed; increase in longevity by 35 percent of 634 mice with cancer tumors, in comparison with the controls; increase in longevity by 23–30 percent of mice pretreated to a field force of 4,200 Oe, and then exposed to deadly gamma radiation; more rapid healing of inflicted wounds—in mice—of those housed in a magnetic field for one month, than of controls housed in dummy magnets. Also "pathologists agreed that fibroblast proliferation and fibrosis are reduced in magnetic fields" (Barnothy, p. 144). How does magnetic energy do this? Barnothy assumes no direct action of it on the biological system, but an indirect action primarily, on the blood. "The effect is apparently related to the *leucocytosis** which follows magnetic treatment" (*ibid.*, p. 138).

Other scientists think that magnetic fields can produce favorable effects upon humans.

> Dr. Robert O. Becker, an orthopedic surgeon on the staff of the Syracuse Veterans Administration hospital and the New York Upstate Medical Center, states that "subtle changes in the intensity of the geomagnetic field may affect the nervous system by altering the body's own electromagnetic field." . . . George N. Chatham of the National Aeronautic and Space Administration, Washington, D.C., related that "careful and precise studies of magnetism and its effect may open up a new approach to biology since the entire body is an electrical organism basically; owing to the characteristic electrical charges and valences of atoms of bio-electric energy in nerves, organs and tissues. Fields of magnetic energy properly applied and directed might therefore affect the electrical response behavior pattern. It has been ascertained that the heart rhythmically beating and motivated by an electrical impulse produces a very minute magnetic field." [*The Helio-magnestat*, pp. 1–2]

All together some four hundred studies on this subject were listed in the index of a single biological journal (*ibid.*, p. 3). In a pioneer article that appeared in the sixties in the *Saturday Review*, scientists Becker, Charles H. Bachman and Howard

---

* Leucocytosis refers to the production of white blood corpuscles, which fight infection and disease.

Friedman probe the effects of magnetism on the human nervous system.

Attempts are now being made by medical innovators to utilize these energies for therapy. Thus the Electronic Medical Foundation has a Depolaray apparatus, which brings a low-frequency alternating magnetic field to bear on affected tissues. It mobilizes about 200 gausses at the center of its face. Its operation "might be likened to demagnetizing a magnet with an alternating magnetic field . . . This field may be applied through a dressing or a plaster cast" (*The Electronic Medical Digest*, Vol. XXXVII, No. 2, [Second Quarter, 1953], p. 10). A quite new device, called a Helio-magnestat, creates a unipolar or unbalanced magnetic field, which causes acceleration of the electron spin. Supposedly it produces some ionization within the muscle cells and tissues with which its electrodes are in contact, and so it can cause vascular dilation or reduction of spasms. The most frequent subjective impression in its use is a feeling of deep relaxation. In *The Cancer Biopathy* Reich refers to an experiment that showed the orgone accumulator affected the N–S orientation of a compass needle. And in *The Discovery of the Orgone* he reported that objects left near the SAPA bions became magnetized. From such observations he concluded that it was likely "that magnetism as such will be shown to be a function of the cosmic orgone energy" (Reich, *The Cancer Biopathy*, p. 108). Perhaps he is overgeneralizing here, but clearly, since the accumulator is made of iron and steel, it would be worthwhile to check to what extent it might produce magnetic effects, or whether orgone is a magnetic-type energy.

Possibly more relevant to the understanding of the orgone is the role of static electricity in healing, which has had a long history. The first medical uses of static electricity go back to Greek and Roman times, when the torpedo fish—with his electric ray—was found to have therapeutic effects on gout and headaches. The first electrotherapist is said to be Jallabert, a German professor of physics who used a static-electricity-creating machine to treat a paralyzed arm (Schwarzschild and Bierman, p. 1467). Three well-known figures, Benjamin Franklin, Marat of the French Revolution and John Wesley, were all interested in the therapeutic effects of static electricity.

According to Sidney Licht, the first systematic utilization of static electricity for medical purposes was in the mid-eighteenth century in Germany. Since then at least a thousand books have been written on electrotherapy (Licht, p. ix). Extravagant claims were made by many therapists, some rather humorous like that of Rousell: "It is especially in the genital organs that electricity is truly marvellous. Impotence disappears, strength and desire of youth return and the man, old before his time . . . can become fifteen years younger" (*ibid.*, p. 19).

With the development of the Wimshurst and Holtz machines, interest burgeoned in the United States so that by the end of the nineteenth century "most [medical] practitioners in America had a static electricity machine in their offices" (*ibid.*, p. 20). The medical apparatus used in the United States was Holtz machines, and these are initially charged by a Wimshurst machine. Their operation differed depending upon the circuit. There is the static bath, the static wave, which delivers a current whose instantaneous value may go as high as 100 amps, or the direct and indirect spark. During treatment, "the patient may complain of a prickly sensation if any metal object such as a ring, a hairpin or a watch is near the skin . . . the hair (too) stands on end . . . . Duration of treatment is usually 20 minutes" (Schwarzschild and Bierman, in Otto Glasser, *Medical Physics*, p. 1491).

Dryness of the atmosphere is essential to the successful operation of static machines. (The orgone accumulator also works much better on dry days and in dry climates.) The beneficial results reported by Schwarzschild and Bierman, research scientists and professors at New York University, are parallel to those mentioned by Halacy. Thus they mention sprains, contusions, bursitis, synovitis, lumbago, and sciatica. "Its ability to cause relaxation of muscle spasm makes it valuable in treatment of conditions where muscle spasm is an associated phenomenon as in arthritis" (*ibid.*, p. 1472). They refer to indolent ulcers as usually treated by the static brush machine. However after the First World War, this kind of therapy began to yield ground, especially in the United States, to pills and then to antibiotics. Thus, by the fifties, texts in physical medicine seldom even mention this therapeutic technique. It may be that both the great wave of interest in this therapeutic method, in the later part of the nineteenth century, and its decline after 1920 represent a failure to explore thoroughly both the real benefits and the limitations of such a treatment process.

It is relevant that certain experiments have suggested that electrostatic fields can enhance the growth of plants and vegetables. These were carried out by Wesley Hicks, an American biometeorologist. Again while Dr. Tromp refuses to assert that the growth effects "were really due to electrostatic fields," he does emphasize "that it is most important that the possible biological effects of electrostatic fields be studied" (Tromp, p. 573). In sum, static electricity apparently possesses therapeutic and growth-enhancing energies which parallel reported powers of the orgone. But the exact connection between static electricity and orgone is unclear, even to such a careful student of Reich as Charles Kelley (Barth, p. 85). Properly qualified scientists would help by carrying out decisive controlled experiments.

Another radiational energy that is known to orthodox science and may throw light on effects of the orgone accumulator are ionized particles, especially those with negative polarity.

Ions are electrically charged atoms, molecules or molecular groups; these charges may be either negative or positive. The existence of ions in the atmosphere is due to the bombardment of molecules in the air by (i) cosmic rays coming from interstellar space, by (ii) solar radiation, by (iii) radioactive materials on and below the surface of the earth, and by (iv) friction and falling water droplets, as in a waterfall. The action of the winds and the fall of barometric pressure cause . . . [these latter] ionized gases to be diffused through the capillaries of the soil into the air. About 60% of the total ionic content of the air near the surface of the earth is due to gases from the soil. Terrestrial radiation from radioactive particles in the air extends a limited distance from the earth. It eventually meets cosmic radiation. "Ionization of the air and therefore its conducting power will be decided by the net effect of the two types of radiation." [Bach, p. 5]

The condition of the atmosphere determines the relative distribution of the various-sized ions. The number of ions in the air varies from 50 to 10,000 for each cubic centimeter of air, depending on many factors. They are higher in number during the day than at night. They are much more numerous on clear, sunshiny days than on foggy and rainy days, and also they are much more numerous in summer than in winter. These characteristics, interestingly enough, also apply to the orgone. The atmosphere contains both negatively and positively charged ions. One or other may predominate in number; on mountaintops, negatively charged ions predominate. With the approach of thunderstorms, the number of positive ions becomes greater. On sunshiny days there is an increase in the number of negatively charged ions, and on cloudy days the number of positive ions is increased.

Experiments in the General Electric laboratories have shown that ions may be produced by electric sparks in the air, by X rays and by radium emanations. Even an open flame in a fireplace ionizes air. With suitable apparatus one may extract either the positive or negative charges and thereby control the kind of ions that remain within an enclosed room. "Normal" air contains both big ions and little ions; it has been discovered that the number of large ions increases after sunset, whereas the number of small ones increases during the early morning hours (*Electronic Medical Digest*, Winter, 1950, pp. 28–32).

A large number of studies of the physiological effects of air ions are summarized in an article by P. Krueger, W. W. Hicks and J. C. Beckett in Tromp's *Medical Biometeorology*. "Thus, compared to controls, negative air ions produced accelerated growth of chicken tissue in vitro. Whereas positive ions in general inhibit growth or activity (in animals and tissue), [and] induce the contracture of smooth muscle and enhance vulnerability to trauma, negative ions accelerate growth and ciliary activity and reverse positive ion effects" (Tromp, p. 362). Later, Krueger reported that negative ions exert a moderate lethal effect on certain bacteria.

An experiment at the Institute of Hygiene, in Prague, with normal males exposed to air ions for one hour three times a week, for eight weeks indicated that positive ions induced a rise in blood pressure, a drop in blood albumin, and an increase in the globulin fraction of the blood. "The only significant effect of negative ions was to raise the blood albumin and to lower the globulin" (*ibid.*, p. 368). Experiments on the effects of ions on animal tracheal strips and subsequently repeated on the tracheal mucosa of rabbits, rats and other mammals concluded "that negatively charged oxygen and positively charged carbon dioxide are the mediators of [air ions'] physiological effects" (*ibid.*, p. 361). In general, positive ions produced an increased irritability in the mucosa. Moreover, subsequent tests showed that an exposure of sixty minutes to air ion concentrations sufficed to initiate changes in the animal's trachea and that these changes, once established, persisted for up to four weeks (*ibid.*, p. 360).

In 1962, Krueger, Hicks and Beckett reported on experiments with positively and negatively ionized atmospheres and their effects on the respiratory systems of selected mammals. They found that positive ions were associated with congestion in the bronchial tubes, interference with the normal peristaltic reflex, a slower beat of the cilia in the windpipe and a reduced flow of mucus. Negative ions had reverse effects, being associated with relaxation and expansion of the bronchia, a restored peristaltic reflex, a higher rate of ciliary beat, and increased flow of mucus. Whereas with positive ions, the cilia beats were 600 a minute, they rose to about 1,000 with negative ionization (Krueger, Hicks and Beckett, eds., *Weather, Climate and the Living Organism*, Amsterdam, Elsevier, 1962).

"Experiments seem to show that it is only the small ions which exert a *direct* physiological influence . . . large ions and particles in the air exert an indirect influence" (Bach, p. 7). One series of interesting tests compared the measured numbers of ions in a room with people's subjective impressions. If there were remarkably few ions in the air, people described the atmosphere as "close." If there were many positive, but few negative ions they described it as "sultry." Where there were many ions of both signs, the air was felt to be "light and fresh." A predominance of negative ions led people to characterize it as "light and cool." Another test showed that in the vicinity of a fountain, there is a surplus of negative ions and people report a peculiar freshness. In nature, negative ions are released from plants, especially from the upper parts; but there are great differences in the plants' capacity to give off ions.

Beginning with Germans and Russians in the thirties and forties, an increasing interest in the therapeutic effects of air ions has developed. A pioneer Russian scientist, Tchizhevsky, claims that "ion therapy can lead to improvement in 85 percent of patients with vascular and cardiac conditions, hypertension, angina, bronchitis, migraine, endocrine disturbances, allergies, burns, bronchial asthma, and stomach ulcers" (Tchizhevsky, *International Conference* . . . , pp. 227–228). In the total literature, which comprises more than two thousand publications, there is some consensus that treatment by negative ionized air usually produced by a variety of unipolar machines has beneficial results on asthma, migraine headache, burns, hypertension, bronchitis, high blood pressure, hay fever,* and vasomotor angina pectoris.

A leading American researcher, Dr. Igho Kornblueth, of Philadelphia, has reported impressive results in the use of negative ionized air with burns. The treatment has been in regular use for some years in the hospital to which he is attached in Philadelphia. "The amount of secretion and the number of infections is substantially reduced. The fetid odor usually accompanying severe burns is completely controlled, and in the great majority of cases a complete cessation of pain is achieved in 10–15 minutes with no need for analgesics. The healing process is also speeded up and rendered more satisfactory" (C. A. Laws and E. Holliday, "Organic Electronics, the First Breakthrough since Antibiotics," p. 21). When we recall the marked results with burns achieved by Reich and others using orgone energy "shooters," one is struck by the possible significance of this parallel.

Some of the literature on the subject is summarized in an article, "The Action of Air Ions on Bacteria," in the *Journal of General Physiology*, November, 1957.

A very interesting study of the subjective and objective effects of air ions on the upper respiratory tract of human beings [indicated] that positive air ions induce pronounced nasal obstruction, headache, dryness of the mucous membranes, husky

---

* In one experiment, reported by Dr. H. Kornblueth in the *American Journal of Physical Medicine*, the hay-fever symptoms returned in two hours when the patients went back to ordinary outdoor air.

voice, and dizziness . . . Negative air ions were much less
active in bringing about any of these phenomena. Similarly
Yaglou and his colleagues have reported irritation of the upper
respiratory tract and headaches accompanying exposure to
positive air ions. Negative ions brought about relaxation and
mild euphoria. [Pp. 363–64]

Christian Bach, a Danish researcher, using a Floraion apparatus in hospitals and in-patient clinics, treated over one
thousand patients for a variety of complaints and 62 percent
gained some favorable effects. Children and persons sensitive
to weather changes were the easiest to treat (Bach, pp. 48–49).
Physiologically, "negatively charged air results in a decrease
in the respiration rate and blood pressure. [It] seems to increase the sensitivity of nerves [and] the regeneration of open
wounds" (Halacy, p. 115). Another interesting finding from
Germany is that the use of negative ionized air seems to lead
"to a decided increase in the hemoglobin, red blood cells and
iron content of the blood" (Electronic Medical Digest, Third
Quarter, 1951, pp. 20–21).

Even more interesting is the observation of M. S. Sinaya
at the Physiological Institute in Leningrad, who found that
under the influence of positive ions, the mobility of the red
blood cells decreased, but increased in a negative ion atmosphere (M. S. Sinaya, "The Influence of Unipolar Air Ions on
the Electrophoretic Mobility of Erythrocytes," International
Journal of Biometeorology, Vol. II, 1967). This takes on increased importance when added to a finding by Katsenovitch
that a negative ion atmosphere results in an intensification
of the activity of the white blood cells, which are responsible
for counteracting infections (R. Katsenovitch, "L'hydroaeroinisation lors du traitement de la phase non-active du rhumatisme").

Another fascinating finding was that certain kinds of ionization may be beneficial to living things up to a certain concentration, after which they can begin to be harmful. This
pattern parallels experiences in the use of the orgone accumulator, as does the following caution:

The susceptibility of human beings to ions is very variable.
Some seem to be extremely susceptible to small alterations in

the ion content; others seem to be practically unaffected by even quite large changes. Ten years' practice of the matter has shown that we find a mixture of quite convincing cases where alleviation of illness, a cure or an improved state of health has demonstrably been achieved by increasing the ion content, whereas other cases give rise to doubts whether there was anything in it at all. In fact, convincing results both for and against the use of ions as a therapeutic and hygienic remedy have been found. [Bach, p. 48]

One is impelled to ask what is it in the unaffected person's constitution that allows him to remain unaffected, or vice versa.

One of the problems mentioned by several experimenters is the length and frequency of the optimum dosage. The usual treatment in Germany in the thirties and forties—some thirty thousand persons were treated there between 1930 and 1945—was twenty daily periods of fifteen to twenty minutes. Overdoses could, it was claimed, cause insomnia. Asthmatics that Bach worked with in Denmark found after considerable experimentation that they got the best results by doubling the daily irradiation. Bach points out that the time needed depends in part on the quantities of ions offered by nature and that these vary with the over-all electric field and the weather. The following conclusion from a speaker at the 1961 International Conference on Air Ionization, in Philadelphia, is noteworthy: "The description and measurement of ions is still so imperfect that it might be compared with men's knowledge of light before the discovery of the prism" (Bach, p. 90). While some advance has been made since then, it is still true that the effects of negative ions are very far from being fully understood.

In the September 1972 issue of *Energy and Character*, editor David Boadella makes some fascinating comments on the possible relationship between ions and the orgone. He notes, to begin, that Friedrich Kraus, a forerunner of Reich, claimed that "the circulation of the body fluids, such as blood, lymph, the digestive juices and the microscopic movements of protoplasm within the cells, depended on the concentrations of ions in the tissues" (Boadella, p. 18). Boadella goes on: "Just as there is a relationship between the flow of ions in the body and the bio-energetic liveliness, so we must assume a relationship between the concentration of ions in the atmosphere (and whether they are positive or negative) and what Reich called the orgonotic tension in the atmosphere" (*ibid.*, p. 19). He quotes studies that indicate a connection between positive ion intake and reduced air intake. He then hints that positive air atmospheres—an excess of positive over negative ions—may be the same as Reich's DOR, since the effects described, dryness, burning and itching of the nose, headache, dry scratchy throat, dry lips, dizziness, difficulty in breathing, etc., are much the same as Reich's effects of DOR. He also quotes a study by Kuster of Frankfurt that found "that negative ions slowed down or arrested the rate of growth of cancer tumors in mice, just as Reich had found a reduction in tumor formation by the use of the orgone accumulator" (*ibid.*, p. 23). Another study, by Puck and Sagik (which he quotes), theorizes that cells and viruses carry the same negative electric charge and that positive ions seem "to inhibit or neutralize cellular resistivity, permitting the virus to enter and pursue its work of destruction" (Puck and Sagik, *Journal of Experimental Medicine*, 97:808–820). From other studies Boadella concludes that air ions can penetrate the skin (like the orgone) and produce functional changes in internal organs. He notes, "This assumption was confirmed by the work of McDonald, who administered ions to rats via the skin, taking care to make sure that they breathed only normal air, and found the predicted changes in blood pressure due to the effects of the two types of ion" (*ibid.*, p. 25).

This leads him to a discussion of a peripheral (skin-located) nervous system. "It consists of a network of fibres in the cutis and epidermis of the skin which are linked to minute autonomic cells. According to H. A. Mayling, the peripheral nervous system may act as the receptor for subtle external stimuli such as changes in the radio-active and electrical properties of the atmosphere. It also seems to be responsible for the tonus of cell fluids and their diffusion" (*ibid.*, p. 25; H. A. Mayling, *Journal of Comparative Neurology*, Vol. 99).

A British engineer and a physician, writing in the same issue of *Energy and Character*, report on conclusions of

German researcher N. Schulz, following many years of investigation, that "the defensive system of the body is dominated by certain of the endocrine glands, notably by the adrenals, which operate through neuro-hormonal mechanisms. He [Schulz] showed that electro-aerosols (air ions) operate directly on the adrenals and central nervous system. These findings are supported by Italian researcher R. Guallierotti ('The Influence of Ionization on Endocrine Glands'), who has demonstrated that negatively ionized air produces definite changes in the functioning of the endocrine system. Two of his several findings are, first, the correlation with the chemical regulator serotonin, and its release from the tissue, a central factor in many migraine conditions; the second is the effect on the cells of the suprarenal glands, which indicates the mechanism underlying the beneficial results [of negative ion atmospheres] obtained with rheumatic disease" (Laws and Holliday, *op. cit.*, p. 21). In summary, a number of independent findings suggest that a negative ion atmosphere is variously therapeutic and that it operates apparently much as the orgone is supposed to. At the same time, research on ions, the blood and the vegetative nervous system throws light on how these electrical charges are related to healthy functioning of the organism. It is very suggestive, too, that both the orgone and negative ions depend for their best results upon specific weather conditions, and on the over-all field force of the individual.

An interesting point has been made by neo-Reichian Lawrence Barth—namely, that an accumulator could capture some negative ions in the following way: The galvanized iron used in making accumulators is composed of iron coated with zinc. Ultraviolet "radiation hitting zinc is known to knock electrons free from it . . . and such free electrons will soon attach themselves to atoms of the air's gases and so create negative ions" (Barth, p. 80). This raises two basic questions: firstly, how does the ultraviolet get from the outer layer to the innermost galvanized layer? Physicists know of no evidence for such conduction. Secondly, to be effective therapeutically, a substantial number of free electrons would have to be knocked off; this would require considerable force in the ultraviolet radiation as it hits the orgone accumulator.

The presence of a zinc layering on the iron in the accumulator raises another possible clue to its therapeutic effects. This hinges upon the fact that zinc has therapeutic properties and these might be picked up and absorbed through the aura of animals and people using an accumulator. First, let us note that zinc, which is coated on the iron in galvanizing, is now known to have therapeutic effects on organisms, when taken internally. Studies in recent years have shown that zinc is "only slightly less abundant than iron in living organisms . . . some enzymes and proteins contain zinc as essential components" (Prasad, p. 69). A respectable hypothesis being tested today is that "the level of zinc in cells controls physiological processes through the formation of zinc enzymes" (*ibid.*, p. 70). Experiments with birds have shown that zinc deficiency is associated with retarded growth and with certain diseases (*ibid.*, pp. 202–4). Studies on many animals have confirmed the necessity of zinc to health (*ibid.*, pp. 250–51).

Prasad summarizes research on zinc's importance to humans as follows:

> Zinc is essential for growth of plants and animals. Testicular atrophy in zinc-deficient animals has been reported. Recently, zinc deficiency has been found to exist in patients from villages in the Middle East who exhibit severe growth retardation and sexual hypofunction. Under the conditions of our experiments, zinc supplementation to such individuals resulted in growth and gonadal development and appearance of secondary sexual characteristics. Detailed investigations for various other trace elements and vitamins failed to reveal any other deficiency except iron, which is not known to cause growth and gonadal disturbances in experimental animals. Some dwarfs who received reagent grade iron only, failed to develop sexually, and their growth rate was less than those who received zinc. Thus, these evidences implicate zinc as being essential for the growth and gonadal function in man. . . . The mechanism by which a deficiency of zinc causes growth and gonadal retardation in animals and men is not [wholly] understood at present. . . . On the other hand, zinc is required for the synthesis of DNA, thus it may be involved in protein metabolism. Deficiency of zinc therefore may affect the growth of bone and testicular function directly. [*Ibid.*, pp. 292–93]

The effect of zinc in triggering a healing process is accepted in the use of zinc compounds such as salves, ointments and lotions. A study to investigate the role of dietary zinc in wound healing showed that "skin lesions developed spontaneously in zinc-deficient animals and heal promptly if adequate [organic] zinc is provided" (*ibid.*, p. 381). Tests with humans were carried out, and it was discovered that "the healing rate of wounds . . . increased with the increasing level of zinc in the [subject's] hair . . . The data . . . demonstrate a good correlation between hair zinc levels and the rate of human wound healing" (*ibid.*, p. 388).

While nothing has been scientifically established about the body's capacity to absorb radiation of metals, two facts suggest that it may be a viable hypothesis. First, we know that all metal radiates, and from studies of the aura, we know it is visually affected by magnets. If we assume that through auric emanations, radiated substances can be carried, in some unknown fashion, to the body—surely not an impossible assumption after what we now know of the body's remarkable sensitivity—then the way is paved to consider the zinc (specifically its frequency of vibration) in the accumulator as the possible source of some of its therapeutic effects.

Finally, in considering this hypothesis it is significant that high concentrations of zinc are toxic and lead to nausea, vomiting and headaches (*ibid.*, p. 412). Typical cases involving humans include food or beverages stored in galvanized tubs (*ibid.*, p. 415). It is suggestive, too, that organs with high concentrations of zinc—for example, the eyeballs—are least susceptible to cancer.

For a variety of reasons ultraviolet radiation seems to be perhaps the closest to the orgone in the range and the character of its therapeutic (and, in heavy doses, dangerous) effects. Thus, both ultraviolet (UV) and orgone can cause tanning of human skin. Reich, for instance, related that in a highly overcharged orgonotic atmosphere he became tanned* even

---

* A single experiment conducted by the writer with an eight-layer orgone blanket failed to produce any signs of skin reddening after two one-hour irradiations in one day of the bare chest. This suggests that, if UV radiation is "produced," its concentration is much less than in ordinary sunlight.

under his clothes, though he was not out in the sun. "I had carried on these experiments during the winter and early spring, had not been in the sun, and yet had a strongly tanned body" (Reich, *Discovery of the Orgone*). Both UV and orgone in moderation create a feeling of vitality and health, whereas in excess both lead to a feeling of overcharge and malaise. Lawrence Barth notes "Reich's Experiment XX showed that orgone has a very direct function in the creation of living organisms. UV by itself, and also electrical sparks [which generate UV radiation] can, in the presence of appropriate chemicals, bring about the synthesis of certain organic molecules that are building blocks of living cells"; this has been shown by the experimental work of S. L. Miller and others, following the theoretical formulations of A. I. Oparin and Harold C. Urey.

> The most common colour of orgone is light blue or light violet. A mercury-vapor UV lamp produces approximately the same colour of visible light along with the UV radiation; and of course UV is next to violet in the electromagnetic spectrum . . . [Again] water absorbs orgone strongly. The same is true of UV. Though his description was not sufficiently clear and specific, Reich said in "Discovery of the Orgone" that a fluorescent glass plate such as is used in a fluoroscope will respond to concentrated orgone and glow under its stimulus. Besides responding to X rays . . . such a screen responds also to UV and, in general, to any electromagnetic radiation including static electricity. In fact, a fluorescent lighting tube depends on this reaction in its ordinary use; it generates UV within the tube, and the reaction of the coating on the inside of the glass to this UV creates the light. [Barth, pp. 78–79]

In short, significant parallels exist between UV radiation and orgone. But subjects of UV radiation fail to report tingling or prickling as with the orgone.

A number of other facts raise the issue of ultraviolet radiation's relation to orgone. First, we recall that Reich's discovery of a radiation from SAPA bions occurred because the bions caused him inflammation of the eyes, or conjunctivitis. Ultraviolet radiation is well known to have similar effects on the human eye (Licht, pp. 272–73). Since the SAPA bions were made from seashore sand, they might concentrate UV radiation, through long exposure to sunlight. Secondly, all the tests on temperature difference (To − T) in the small-orgone accumulator show it is highest on sunny days around midday, which fits neatly with the times of maximum UV radiation in the atmosphere. Thirdly, as with the orgone, there are among humans "large individual and conditional variations in susceptibility to UV radiation" (*ibid.*, p. 267). Hair color, age, sex and season affect an individual's photosensitivity to it. As with the orgone, repeated exposure to UV radiation diminishes the original sensitivity—namely, reddening of the skin. Again UV radiation, just like the orgone, is described by reputable scientists as energizing, building resistance to disease, able to kill bacteria and viruses, healing wounds, increasing germination and the growth of seeds (wavelengths over 300 $\mu v$), improving muscle tonus, dilating blood vessels and causing nerve excitation. Some of the bacteria found susceptible to UV radiation are tuberculosis, staphylococcus, streptococcus, anthracis, E coli, alpha and beta pestis (*ibid.*, p. 189).

In therapeutic benefits so much has been claimed for UV radiation that one again thinks of the orgone. Favorable therapeutic results range from skin conditions like psoriasis, boils, eczema, to TB, to ulcers and rheumatic conditions, to asthma, constipation, catarrh, to anemia, arteriosclerosis, hypertension, varicose veins, bronchitis, chorea, neurasthenia, sciatica, gout, arthritis, nephritis, vaginitis, hay fever, dental caries, colds, rickets and even pyorrhea (*ibid.*, pp. 195, 332).

In addition, the effectiveness of atmospheric UV is related, like the orgone, to the character of the cloud layers—for instance, thick dark clouds reduce the intensity of UV radiation. Other factors which affect UV radiation are the concentration of ozone, Rayleigh's scattering, aerosols and smog (Urbach, p. 429). But Urbach admits that "the attenuation of UV in the atmosphere is a very complex phenomenon and many details remain to be worked out" (*ibid.*, p. 333). UV radiation supposedly makes for an increase of the leucocytes, of the erythrocytes and hemoglobin values, a decrease in blood sugar, an increase in protein metabolism, an increase in liver and muscle glycogen, and a drop in blood lactic acid—effects which very much resemble those caused by insulin. Possibly some or all of the above physiological effects result from UV's tendency to make the subject feel warm and relaxed.

Most of these effects have also been claimed for orgone. Like orgone, UV leads to a decrease in blood pressure and a change in cardiac output and cardiac rate (*ibid.*, p. 280). Gastric secretions also increase, and there are alterations in the functions of the adrenals, and the sex organs. There is a "considerable literature claiming remarkable improvement in performance of athletes after repeated and consistent UV exposure" (*ibid.*, p. 281). The following conclusion from Licht also reminds one strongly of insights about the orgone: "The magnitude of the effects depends on the strength and duration (as well as the wavelength) of the irradiation . . ." (*ibid.*, p. 282). Finally, Barth points out that the mitogenetic rays given off by many kinds of organisms are of the general UV wavelength, but somewhat shorter than the sun's UV wavelength, which means they would be more lethal germicidally than the sun's form of UV radiation. Urbach adds: "There are still a great number of medical phenomena and chemical reactions which are caused by light or UV radiation, of which we have but little knowledge" (Urbach, p. 105).*

However, at least two aspects of ultraviolet radiation seem to contradict any clear identity with the orgone. First, as is well known, repeated and heavy UV irradiation can and does lead to skin cancer. "Massive doses can trigger off a cancer effect" (*ibid.*, p. 88). This conflicts with orgone's apparent ability to reduce cancerous growths. Secondly, simultaneous mixing of UV with X-ray radiation results in a reduction of X-ray effects (Licht, p. 251), whereas when X rays are activated near orgone, it is supposed to produce DOR effects or dangerous radioactivity.

These numerous parallels in the effects of orgone and UV radiation raise questions about the present theory of the orgone. One is led to ask, Is orgone energy a fancy name for UV radiation, or closely related to it? To date, no Reichians or neo-Reichians have tried to systematically explore these parallels, but it clearly needs to be done. Owing to medicine's con-

---

* Urbach also notes: "It is hoped that the use of new radiation sources (new lamps, etc.) may lead to a large number of new findings."

centration on antibiotic drugs to control many conditions earlier treated by UV radiation, there is a dearth of recent research on UV's therapeutic values. Numerous suggestive areas of research on UV's effects still remain to be explored. Among these are the differential therapeutic and germicidal effects of UV radiation of differing wavelengths (*ibid.*, p. 244).

Perhaps burgeoning interest in the orgone theory will spill over and encourage more research into therapeutic uses of UV radiation. Such investigation would help clarify the relationship of these two sun-based energies. Meantime, it is clear that since orthodox science accepts the therapeutic value of various energies—magnetic, electrical and ultraviolet—though in each case in a limited way and dependent on specific devices, it may shortly begin to look into Reich's claims for orgone energy.

We conclude this survey of orthodox science's understanding of invisible life-enhancing energies by itemizing some little-known facts about the atmosphere, facts that bear on the notion of atmospheric orgone. We know man's behavior is somehow affected by solar flares, sunspots, the earth's geomagnetic radiation, cosmic rays and atmospheric electricity. Taken together with ultraviolet radiation, which persists in some strength even on cloudy days, it is evident that human organisms are constantly bathed in a sea of invisible energies. The basic studies of these energies, medical biometeorology and biorhythms, are only gradually securing much attention in North America. These sciences reveal something of the complexity of the invisible energies incessantly bathing earth's organisms. Specifically with respect to pulsating electrostatic fields or electromagnetic fields in the atmosphere, Tromp notes that "many experiments support the view that [such] strong . . . fields . . . within certain frequency intervals may have important biological effects on the nervous system of man and animals" (Tromp, *Medical Biometeorology*, p. 455). One result of the variety of electromagnetic waves being shot at us from outer space along with cosmic rays is that there is a potent electric field in the atmosphere. A constant electric voltage exists, varying with the height above ground; for instance, at six feet it is about 200 volts. Between the ionosphere (at fifty kilometers) and the earth, the potential is 360,000 volts

(Halacy, *Radiation, Magnetism and Living Things,* p. 17); but it has little current.

It is also significant that since the earth is a huge magnet, it is surrounded by a colossal magnetic field of force. Indeed, there is a solar wind—a cloud of charged particles moving at supersonic speeds—which when it strikes the earth's geomagnetic field "gives rise to great tongues and sheets of electromagnetic radiation" (*ibid.*). At a height of about 80 miles an auroral electrojet—a great current of charged electricity— is thought to flow, which near the magnetic poles has a current estimated at perhaps 500,000 amperes (*ibid.*, pp. 18–19). Tides in the charged particles surrounding the earth also affect the surging currents of electricity in the atmosphere. So there are electric currents near the earth's surface. There is also a current moving in both directions vertically. Tromp notes: "The electric field of the atmosphere and the negatively charged earth's crust, together create an air-earth current which continuously neutralizes the negative earth charge with the positive charge of the atmosphere" (Tromp, p. 75).

When Reich asserts the presence of an orgone (life-energy) envelope around the earth, he may be right or may be simply drawing attention to the evidence of various electrical fields which affect the atmosphere and, through it, human feelings and responses. This chapter helps to underline, on the one hand, the plausibility that electrical or partly electrical energies play a significant role in human reactions, and on the other hand, the complexity of the energy system in which the earth is enveloped. Whether the orgone is a separate force alongside known atmospheric energies or an integrated theoretical notion pointing to a complicated collection of forces only further research can tell.

*Chapter Fourteen*

COSMIC ENERGIES AND THE ORGONE

The plausibility of certain aspects of the orgone theory requires evidence for (a) the existence of invisible micro "waves," or energies, having a nonterrestrial source, and (b) an organism whose state of dynamic equilibrium is changed by these or some unknown forces. It is also related to notions of cosmic equilibria to which the organism (man), the weather and far-ranging energy forces are in intimate association. In effect, it claims that the cosmos and all its parts, since animated by a single, primordial cosmic energy, are closely interdependent. Thus, in *Cosmic Superimposition*, Reich says: "From the orgone ocean all being, physical as well as emotional, emerges. Man . . . is, together with all living beings, a bit of especially varied and organized cosmic orgone energy" (*Cosmic Superimposition*, p. 12).

Two new, perhaps border fields of regular science, which have since the late fifties reached a significant stage of development, add a fascinating array of confirming data to assumptions of, or elements in, Reich's orgone theory. These two fields are biometeorology and the study of biological rhythms. The first looks into the influence of atmospheric and cosmic conditions on life, and the second investigates the meaning and relevance of biological rhythms. Two international societies unite these disciplines—the International

Society of Biometeorology and the Society for Biological Rhythm. Until twenty years ago, inquiries into weather effects or lunar and planetary influences were considered the domain of vague, unrealistic quasi scientists and often the privileged hunting ground of charlatans. Such for example were studies like astrology or attempts to link the moon to accelerated plant growth.

Slowly scientists moved into these areas. One of the first was the Swedish physicist Svante Arrhenius—winner of the Nobel prize in physics—who studied menstrual cycles (of 11,807 persons) and "concluded that their frequency during the waxing moon was higher than during the waning moon, reaching a peak on the new moon's eve" (M. Gauquelin, *The Cosmic Clocks*, p. 167). In recent years, as meteorology has progressed significantly, some of the early hostility of some established scientists to these new disciplines has abated. Thus, according to A. Sollberger, a respected American researcher, "biological rhythm research [originating in Germany and Scandinavia] has finally become recognized as an important interdisciplinary branch of science. . . . A not unimportant impetus was derived from space medicine with the problem of how the body rhythms are controlled from without and what happens if this influence vanishes" (A. Sollberger, *Biological Rhythm Research*, p. 308).

There is the growing assumption, due to space probes, that interstellar space is not empty, but is filled with a medium, now often called *plasma* (an electrified gas), a fourth kind of matter. While some science writers see intergalactic space filled with magnetic fields "holding" the galaxies in shape, others are content to see space as interlaced with many force fields.* Either notion fits with Reich's denunciation of an earlier idea of "empty" interplanetary space. He claimed that it is filled with living energy, the cosmic orgone. "Space is not empty but is filled in a continuous manner, without gaps" (*Cosmic Superimposition*, p. 26). He also claimed that the northern lights are an orgone-energy irruption. "The aurora borealis, or 'northern lights,' are the effect of orgonotic lumi-

---

* Professor Francis Bitter, chief planner for the National Magnet Laboratory, writing in *Saturday Review*, February 3, 1962, says: "Today ['the ether'] is thought of as containing a variety of 'fields.' "

nation at the outer fringes of the orgone-energy envelope of the earth planet" (*ibid.*, p. 65). After noting the bluish color of the northern lights, Reich elaborates his claim for its orgone character as follows:

> The *movement* of the streamer type of the aurora borealis is of a slow, undulatory, at times pulsatory and wavelike nature. Slow expansion and contraction as well as fast-moving protrusions as in the protoplasm of amebae is characteristic of the aurora. This motion is of the same kind that can be obtained in highly orgone-charged argon gas tubes, through excitation by a moving orgone energy field derived from the body or the hair. Furthermore, some aurora movements have a pushing or searching expression. This, of course, does *not* mean that these phenomena are life expressions. It only means that the same energy that constitutes the bio-energetic movements of pushing and searching are also present in the non-living realm of nature. [*Ibid.*, pp. 66–67]

Further, Reich reasons, from observations of the corona of the "lights," that it results from two orgone-energy streams meeting at a sixty-two-degree angle (*ibid.*, p. 81).

In orthodox science, the aurora borealis is "generally attributed to neutral streams of charged particles, protons and neutrons, emitted from active regions of the solar surface. These gas clouds reach the earth in approximately one day and, under the influence of the geomagnetic field, are deflected toward the auroral zones" (J. A. Ratcliffe, *The Physics of the Upper Atmosphere*, p. 214). There appears to be a positive relationship between auroras and magnetic activity. Specifically, "Clayton has made a statistical study of the association between aurora and sunspots. He found that most auroras occur about one and a half 24-hour days after a spot or group of spots crosses the central meridian (of the sun); that the area of the spots preceding brilliant aurora is much larger than the area of spots preceding moderate aurora; and that the effectiveness of a spot of given size on the central meridian is a decreasing function of its distance from the center of the disk" (*ibid.*, p. 288). Besides this, orthodox science sees the sun as mainly responsible for the negative and positive ions in the earth's atmosphere. Further, Sollberger in his

*Biological Rhythm Research* points out, "The showers of radiation from solar eruptions set up reactions in the outer atmosphere affecting the degree of ionization and electromagnetic phenomena" (A. Sollberger, p. 157). Thus, various electrical energies have a solar origin as does Reich's orgone. Critics are quick to point out, however, that it adds nothing to scientific knowledge to claim the sun as the source of vast cosmic energies. To be really useful, a theory has to be more specific and experimental. This is where Reich was apparently weak.

In *Cosmic Superimposition,* Reich speaks of a cosmic orgone envelope around our planet and we find some apparent confirmation of this idea in the following discovery: "Photographs brought back by our astronauts show that earth is enveloped in a brilliant deep blue" (Heinz Haber, *Our Blue Planet* [New York: Scribner, 1969], dust jacket). Also, "when Mars is in opposition [that is, when earth holds the same position in relation to Mars as the moon does in relation to earth during a solar eclipse] the *blue* screen which hides all the planet's details in some filters suddenly begins to disappear" (Gauquelin, *Cosmic Clocks,* p. 175). On the other hand, orthodox science claims the blueness of the sky is an artifact, and attributes it to Rayleigh scattering of sunlight from the molecules of the air. To claim it is evidence for cosmic orgone energy coming from the sun is to disregard a settled finding of science in favor of an intuitive hunch or imaginative leap. Another possibility is that what Reich called the orgone envelope may be the earth's magnetosphere, which protects the earth by trapping electrical particles up to a distance of 60,000 miles.

While it is not germane to the basic intention of this chapter, it is interesting to note Reich's major assumptions in *Cosmic Superimposition* regarding a wide variety of celestial phenomena.

We must, accordingly, assume that:
1.  Each planet possesses a disc-like orgone energy envelope which rotates faster than the globe.
2.  All planets swing in a common galactic orgone stream, coordinated in time and plane of general motion.
3.  Celestial functions such as sunspot cycles, aurora borealis,

hurricanes, tides, major weather phenomena, etc., are immediate expressions of an interplay of two or more cosmic
orgone energy streams. [*Cosmic Superimposition*, p. 96]

Some of these statements fit in with recent findings, but
others go far beyond conventional science. The two Van
Allen radiational belts of high-energy protons and neutrons
may conceivably be labeled an energy envelope, but so far as is
known their radiation is hostile to human life. As for a
common galactic stream, science today knows only of streams
of charged particles emanating from the sun and the planets
—and the other stars—but there is no evidence for a *common*
galactic stream. To ascribe, as in point three, so many different celestial functions to the interplay of two or more
cosmic orgone-energy streams, is really to simplify a lot of
diffuse data. The grain of truth here, however, is that it
seems as if corpuscular energy (protons, neutrons and electrons) from the sun, when it meets the earth's ionosphere
and our geomagnetic field, may well effect hurricanes and
other weather disturbances, as well as the aurora borealis.
The big question that has to be raised is, Are orgone streams
quite different from these interacting radiational and magnetic
fields, or do the latter "equal" orgone, or is it simply misread-
ing the facts of upper-atmosphere physics to speak of orgone
energy being there at all. One possibility is that, writing in
the very early fifties, when our information about the upper
atmosphere was skimpier than now, Reich may have been
reaching out for what is now understood in a more systematic
and precise fashion.

Reich repeatedly stressed how water has a "hunger" for
orgone–that is, water seeks orgone. We are now informed
by these biometeorological scientists (a) that water has an
unusual and very sensitive structure (especially sensitive to
electromagnetic fields) (Gauquelin, *Cosmic Clocks*, p. 220),
and (b) that it is probably through the water in organisms of
all sizes up to man that they are sensitive to extremely delicate
cosmic influences. Thus, "space and cosmic forces affect us
through the intermediary of water," claims biometeorologist
Gauquelin (*Scientific Basis of Astrology*, p. 220). And its
structure is especially "unstable and easily influenced at a
temperature of between 81 degrees F and 108 degrees F (*ibid.*,

p. 221)—that is, the range of bodily temperatures! This may confirm the great stress Reich placed on water as an attractor for biological energies.

We can now look closely, first at some of the scientific starting points of these two fledgling disciplines and then at some provocative findings or claims. While not new to astronomers, perhaps to lay people it will come as a surprise that the energy outpourings of the sun are quite inconstant, although this variability exhibits some periodic constants. For example, the phenomenon of sunspots, which emit powerful energies, besides solar eruptions or storms, means that our parent sun shoots out varying amounts of energy over time.

These eruptions can produce communication blackouts, magnetic storms and auroral displays; they also produce violent changes in the intensity of the Van Allen belts. Increases of surface brightness on the sun are known as *faculae* ("torches"), *flocculi* ("flakes") and *flares*, which are especially intense. These all appear mainly in the vicinity of sunspots. Flares are of varying size, last for only a few minutes as a rule, and are probably electrical in origin. As many as ten may occur on any day. The more intense ones are called "outbursts," and will often increase the region's brightness tenfold. Distinct from flares are eruptive prominences that rise high in the sun's corona with velocities of hundreds of kilometers per second. Then there are solar storms or disturbances lasting for several hours or for days. They occur in large sunspots and in the corona and affect the radio wave band.

Sunspots appear to be deep, dark regions, or vertices, on the sun's surface, but their temperature is only a couple of thousand degrees less than the rest of its surface. They possess strong magnetic fields greatly exceeding that of the earth. They usually occur in pairs, one positively and one negatively charged, and these are linked by lines of magnetic force. Scientists have uncovered an eleven-year cycle, or periodicity, in the appearance of sunspots. In this period they rise to a maximum number. Research has indicated that trees grow faster in the years of these sunspot maximums, and wheat yields are significantly higher. In these years also occur heavier rainfalls that raise the levels of big lakes, and more storms and hurricanes in tropical regions on earth.

"Meteorological changes also are . . . related to geomagnetic storms, which in turn depend on events taking place in the sun" (Abetti, p. 159). In general, "variations in the number of sunspots and also in the number of flares . . . and other phenomena associated with them affect very strongly the upper layers of the terrestrial atmosphere" (Gamow, p. 103). At these times the sun ejects more charged particles, mainly protons and neutrons, and the magnetic fields associated with the sunspots accelerate their velocities. Violent solar storms also produce strong ultra-violet emissions and considerable corpuscular radiation. "Both of these, particularly the former, may be the direct or indirect cause of increased plant growth" (Abetti, p. 159).

Another basic consideration is that from our space ship earth's standpoint, sun energy affects us differentially at different times owing to the complex way the sun moves through the heavens. First, let us remember the sun keeps spinning around its own axis, and that it turns different sides of itself (with their eruptions) to face our earth in its regular twenty-seven-day cycle. Further, while the earth moves around the sun at eighteen miles per second, the latter is also moving toward the constellation of Hercules at twelve miles per second. All this means the earth describes a spiral or corkscrew type of trajectory and in so doing, it transects different galactic force fields, as it moves along (Gauquelin, *Scientific Basis of Astrology*, p. 218). In short, the earth is continually exposed to varying amounts of sun energy, to magnetic force fields and to cosmic rays.

Besides the sun, most planets in our solar family, and the moon, continually send out electromagnetic waves. These electromagnetic influences, especially those from the sun, seem to alter terrestrial magnetic fields; secondary effects necessarily follow. The earth is also linked electromagnetically with the moon, by virtue of its magnetospheric tail, which extends beyond the distance between the earth and the moon. "According to geophysicist E. G. Bowen, the magnetospheric tails of the other planets extend the same distance into space" (Gauquelin, *The Cosmic Clocks*, p. 113). Precisely what influences are carried or mediated by such magnetic "tails" from the planets have yet to be fully uncovered.

ORGONE, REICH AND EROS

Alongside this information, we must place the new observation that even some of the tiniest of earth's organisms are so sensitive that they respond to microwaves from far distant sources. One pioneer in this field is Professor Frank A. Brown of Northwestern University. Through many years of research, he has shown that organisms like potatoes, oysters, crabs and seaweed respond in their metabolism to fluctuations in geomagnetic fields, or to lunar influences (F. A. Brown, *Biological Clocks*). It is demonstrated, Professor Brown writes,

> that animals such as snails and planarians are able to resolve differences in electrostatic fields of the order of strength of those to which they are steadily subjected in nature. The sensitivity which has been proven indicates the living thing to have more than 100 times the sensitivity which would be required, for instance, to perceive the electrical field created by a thundercloud rising miles away on the horizon.

H. L. König, of Munich, has shown that "waves of very low frequency (such as are released by solar eruptions) . . . affect the sprouting of wheat, the growth of bacteria and the activity of insects" (*ibid.*, p. 142). American scientist Dr. Robert O. Becker has established that "subtle changes in the intensity of the geomagnetic field can affect the nervous system by altering the living body's own electromagnetic field" (*Newsweek*, May 13, 1963). The work of Ravitz described in Chapter Five also supports this finding.

Intriguing revelations on potential cosmic influences have resulted from tests of the sensitivity of inorganic colloids carried out in painstaking researches during the fifties by Professor G. Piccardi, of Florence, Italy. The problem was to explain why the chemical compound trichloride of bismuth, when poured into distilled water, precipitates at a varying rather than a constant speed. On the assumption that solar eruptions were a factor, a controlled experiment was set up, duplicated by a researcher from the Electrochemical Institute of Brussels and repeated thousands of times. Results showed that the unshielded (by metal) compound (as compared to the sheltered control) varied in its precipitation rate in strict accordance with solar eruptions as charted by astronomers (G. Piccardi, p. 86). It would seem, then, that even inorganic compounds are sensitive to fluctuations in solar energy output.

From these scientific beginnings a number of researchers have in recent years turned up correlations that stagger the conventional human imagination. (And while these correlations may seem to give some small comfort to the pseudo science of astrology, none of their discoverers is interested in supporting or confirming that occult study.) Thus, Piccardi, at his Center for the Study of Fluctuating Phenomena, in Florence, has distinguished three types of cosmic fluctuations: short-term variations, such as sudden solar eruptions or great showers of cosmic rays; variations according to the sun's twenty-seven-day cycle and those following its eleven-year sunspot cycle. A lunar effect was also established by changes noted in the speed of solidification of the chemical naphthalene at the time of the full moon (Gauquelin, *The Cosmic Clocks*, p. 217). Other investigators have linked the occurrence of typhoons and changes in barometric pressure to solar eruptions. It is interesting that Reich claimed that typhoons and hurricanes have an orgonomic base (*Cosmic Superimposition*, p. 87). In Japan another painstaking scientist, Professor M. Takata, discovered what became known as the Takata effect —namely, that variations in solar energy cause irregularities in the flocculation index of male blood serum (Gauquelin, *Scientific Basis of Astrology*, pp. 224 ff.). Then a Russian investigator, N. Schulz, in 1954, in the Report of the Academy of Sciences of the U.S.S.R., revealed how a series of 120,000 measurements in the Crimea indicated that the number of white cells in the blood of healthy subjects varied with the number of sunspots. In 1957 he claimed to find an exact monthly correlation between the leucocyte content in the blood and the sun's activity (*ibid.*, p. 228).

This Russian investigator concludes that "Variations in the number of sunspots have an effect on all life on earth." Thus, Sollberger, writing cautiously in 1965, observed: "With regard to the delicate ionic equilibrium in the body, it is not unreasonable to expect an external influence on them" (Sollberger, *op. cit.*, p. 146).

A long series of other intriguing correlations have been claimed. Dr. Sardou, of France, found morbidity positively associated in France with sunspot activity, especially sudden deaths (R. Tocquet, *Cycles et rythmes*). Dr. Lingemann, in West Germany, related solar activity to pulmonary hemorrhage; various other investigators agree "that sunspot activity spells danger for those with pulmonary disease" (Gauquelin, *The Cosmic Clocks*, p. 161). Convulsions during pregnancy (called *eclampsia*) were found to mount on days of special solar activity (*ibid.*), as do cardiovascular attacks and blood clots (*ibid.*, p. 160). Accidents to car drivers and to miners "increase ten percent on days following solar eruptions" (*ibid.*, p. 162). Suicide has been similarly linked by German researchers (Sollberger, p. 168). Various emotional disturbances seem to follow the same pattern: American observers correlated psychiatric hospital admissions with days of strong magnetic disturbances, which are usually associated with sunspots (*ibid.*, p. 163).

As early as 1930, the French scientist Father Moreux wrote:

> The electric flow emanating from the sun affects our nervous systems, and I have often observed that many people, and especially children, are more irritable during times of excessive solar activity. The number of punishments in schools is always higher during magnetic disturbances caused by turmoils on the solar surface. These unconscious influences are expressed, in certain cases, by attacks of nervousness causing complex effects on morbid natures, such as dejection, attacks of gout or rheumatism, headaches, neuralgia, and even temper tantrums. [Gauquelin, *The Scientific Basis of Astrology*, pp. 208–9]

It is rather fascinating that from an orgone-theory standpoint most of the complaints mentioned by Moreux reflect either overcharging—by orgone—of the organism, or the stasis of an increased charge in some part of the body due to muscular armoring. Most, if not all, of the previously mentioned correlations—for example, dejection or gout—would be explainable in terms of a tight armored organism being subjected to a sudden burst of extra energy or orgone or heightened excitation due to the effect of solar changes on its auric field. In short, superficially at least, Reich's concept of a sun-based orgone energy seems capable of explicating all these otherwise inexplicable correlations.

Lunar changes associated with the full moon also precipitate sudden energy increases and are apparently linked to a variety of physiological effects. It is well known that the number of births goes up at the time of the full moon. But scientists have now discovered that a certain shellfish (*kammuschel*) produces its eggs when the moon is full and that California cuttlefish always spawn three days after the full moon (*ibid.*, p. 190). Professor Brown, of Northwestern, found that the fiddler crab changes its color in relation to the position of the moon, and that oysters responded similarly (*ibid.*, p. 192). "Lunar-day variations in $O_2$ metabolism were found in seaweeds, carrots, potatoes, worms and snails" (Sollberger, p. 299). Rats inside a closed room showed peaks of activity clearly related to the position of the moon. Other investigators have established positive relationships between the full moon and outbreaks of violence and emotional disturbance among the psychiatrically hospitalized (Gauquelin, *The Cosmic Clocks*, p. 164). A German physician, Dr. H. Heckert, claims to have established "significant correlations between the lunar phases and a variety of biological phenomena, such as the number of deaths, the . . . occurrence of cases of pneumonia and the amount of uric acid in the blood" (*ibid.*, p. 166). And a Florida doctor found excessive postoperative bleeding, in a study of over a thousand cases, correlated preponderantly with the times of the full moon (*ibid.*, p. 167). Sollberger cautiously writes that "since the terrestrial magnetic field possesses rhythmic components derived from solar and lunar influences it may perhaps act as a carrier for these influences" (Sollberger, p. 147). He adds, "How these various synchronizers [factors] could act upon the organism we do not know . . . It has to be at the cellular level. A direct effect on the delicate ionic membrane equilibria, the orientation of free radicals or the colloid state has been advocated. In higher animals, this action may be on the nerve cells (H. Laurell, *Cosmic Influences upon the Human Nervous System*). It has also been suggested that the very basis of living matter, the semicrystalline structure of the cell water may be affected, "with secondary influences upon the proteins" (*ibid.*, p. 148).

Many of these correlations like those with sunspot activity

could be explicated by the orgone theory—that is, as un-
expected energy bursts causing disturbances to body or
psyche. Another fascinating study using data from the United
States, New Zealand and Australia found that the days of
heaviest rainfall "occurred immediately following the new and
full moon" (Gauquelin, *Cosmic Clocks*, p. 98). In short, the
moon, whose influence upon the tides was not established
scientifically until Newton's day, may have numerous other
effects on the life and well-being of this planet. Reich did not
specifically deal with this, but insofar as his orgone theory
links cosmic sun-based energy to the life and activity of all
earth's organisms, it may well possess some interpretive value,
in the face of these emerging cosmic relationships.

Actually, much of what these various scientists are turning
up seems to be but an elaboration of ancient, often long-
forgotten insights. For instance, in ancient Chinese thinking
the cosmos in its entirety was seen as a living, real and
sacred organism (G. Couteneau). In Taoism, the Tao is conceived as a vital energy that makes the crops grow, as well as
ordering the path of the stars. And as Gauquelin notes, "in
China, as in India and some other cultures as well, the air
was believed to be filled with the grains of life descending
from heaven" (Gauquelin, *The Cosmic Clocks*, p. 12). In fact,
in almost all the early religions, especially animism, a central
belief was the harmony of life both personal and social with
the cosmos, including space and the seasons. Thus man was
seen as a microcosm, the heavens the macrocosm; between
them ran strong currents of sympathy.

In his *Tertius Interveniens*, the great Kepler warned his
colleagues, "who, while rightly rejecting the superstitions of
the astrologers, ought not to throw out the baby with the
bath water" (J. Kepler, VI, p. 145). For Kepler believed that
"everything that is or happens in the visible sky is felt in some
hidden fashion by earth and nature" (J. Kepler, *De Stella nova
in pede Serpentarii*, Vol. I, p. 147). It is understandable that
along with Giordano Bruno, Reich found much to admire in
Kepler. Writing about these subtle influences on biological
organisms, Sollberger provides a reasoned and thoughtful
perspective: "a short contemplation shows that the scope of
these problems is almost frightening to the modern mind.
There are the holistic concepts of all things' being interrelated.

Gravitation and magnetism are dependent on the position of distant celestial objects. Why, that's almost astrology . . . Clearly, we must be careful in accepting such statements but also in rejecting them because of the negative associations they carry in our minds. The problem constitutes perhaps one of the most fascinating challenges to the biological scientist of today" (Sollberger, p. 149).

In our day Reich may be seen as one of the few scientists to take seriously man and animal's rootedness in the whole cosmic order. In *Cosmic Superimposition*, he argues that the genital embrace of animals and humans duplicates the physical juxtaposition of the galaxies—in short, a drawing-together of energy systems in superimposition is written into the heavenly configurations. One may call this extreme and subjective, but at least it clearly symbolizes the similarity of procreative activity, man's basic act of creating and energizing, with the universal order. Piccardi puts it this way:

Our research . . . will be able to show how man is linked to his environment with bonds which escape his immediate apprehension, if not awareness. Only by understanding the mechanism which connects him to the earth and the sky will man be able to understand better his physical and psychic position in the universe today. [Piccardi, quoted in Gauquelin, *The Scientific Basis of Astrology*, p. 231]

The geological-historical basis for these bonds and their timeless significance is admirably expressed by Professor Frank Brown:

Man did not arise on planet earth abruptly and *de novo*. He came gradually to what he is through an orderly transformation, beginning probably from chance chemical complexes in the warm primeval oceans, as first life arose as primordial bits of slime. It is only fitting, therefore, that we seek man's cosmic roots in his long evolutionary past. Biological clocks geared to the major cosmic periodicities appear omnipresent in living creatures. Their occurrence ranges from unicellular forms and flowering plants to mammals, including man. This indicates the ancient, deep-seated character of the relationship of man to the Universe. [Gauquelin, *Cosmic Clocks*, p. xi]

With this type of thinking Reich's orgone theory is in fundamental agreement.

*Chapter Fifteen*

# Russian Science and Bioplasma

Confirmation or support for various aspects of the orgone theory has come in the last few years from a surprising source—namely, Russian and Soviet-bloc scientists exploring ESP.* Both Russian and other Communist governments are pouring money into this area, possibly hoping to use their findings to some military advantage. Some of their findings and technical inventions seem very close to a Reichian accumulator—certainly their theories, as reported by Ostrander and Schroeder in *Psychic Discoveries Behind the Iron Curtain*, reflect a fundamentally similar view of energy flows.

One major breakthrough has been in the detection and also the photography of the body's force field. At the Laboratory for Biological Cybernetics in the University of Leningrad's physiology department, a research group under Dr. Pavel Gulyaren charts the force field with sensitive high-resistance detection electrodes. Gulyaren's "electro-auragram" device reveals much about the energy state of the organism, showing that the emanations change according to health, mood and character. "In 1968, Doctors V. Inyushin, V. Grishchenko, N. Vorobev, N. Shouiski, N. Fedorova and F. Gibadulin announced their discovery: All living things—plants, animals and humans—not only have a physical body made of atoms

* By 1969 there were seventy Russian scientific publications dealing with parapsychology, or ESP.

and molecules, *but also a counterpart body of energy.* They called it 'The Biological Plasma Body' " (Ostrander and Schroeder, p. 213). They described their research in a long paper entitled "The Biological Essence of the Kirlian Effect."

The Kirlian effect refers to a young electronics wizard from Krasnodar named Semyon Kirlian, who devised a machine to photograph bodily emanations. It is a high-frequency spark generator, or oscillator, generating 75,000 to 200,000 electrical oscillations per second. It can be connected to clamps, plates, optical instruments or even an electron microscope. "The object to be investigated (finger, leaf, etc.) is inserted between the clamps along with photo paper. The generator is switched on, and a high-frequency field is created between the clamps and apparently causes the object to radiate some sort of bio-luminescence onto the photo paper" (*ibid.*, p. 199). Leaves and fingers were photographed. The finger, for example, showed points, flares, craters of light, glowing lights. The human hand gave off golds and blues, and a kind of firework display, evidence of flows of energy. Withered or dying leaves gave off no flares or sparks, their dim "clouds" barely moved. In sum, the photographs showed "intense, dynamic energy in the healthy leaf, less in the withered leaf, nothing in the dead leaf" (*ibid.*, p. 200).

By 1949, the Kirlians had a whole array of instruments through which to examine the play of high-frequency currents on humans, plants, and animals, as well as inanimate matter. They felt by then they had perfected the technique enough to show their results to biologists, physiologists, botanists, and other scientific specialists. Soon the greats of the Soviet scientific world began to trek to Krasnodar. . . . There were members of the Academy of Science, ministers of the government. Over some thirteen years, there were hundreds of visitors. Biophysicists, doctors, biochemists, electronics experts, criminology specialists, all appeared at the door of the little one-storied, prerevolutionary wooden house on Kirov Street in Krasnodar. [*Ibid.*, p. 201]

Subsequent tests showed that, as students of the aura had maintained, aura photographs revealed surprising things. For instance, they showed up disease, so one could predict the approaching death of a poisoned leaf. By diagnosing plant disease before it struck, they provided a way of saving agri-cultural crops. Nervous disorders and excitement in humans were also mirrored in the photographs. Besides approaching illness, states of mind, thoughts and fatigue were revealed by the aura's photographs—just as clairvoyants would expect!

After over twenty years of painstaking lonely research the husband-wife Kirlian team received Soviet government recog-nition: "Suddenly the Kirlians were given a pension, a new apartment in a pleasant new district in Krasnodar and a spe-cially equipped lab. Suddenly, full-scale scientific research began on Kirlian photography in institutes, labs and univer-sities all over the U.S.S.R. In 1962 the *Soviet Union Maga-zine* reported that entire scientific research establishments had been set working on the Kirlian phenomenon" (*ibid.*, p. 209). Articles and books about their work emerged from Russian scientific establishments in the late sixties.

Biologists at the State University of Kazakhstan recently described their conclusions about the living energy body. They have declared the bioluminescence revealed in the Kirlian photographs as caused not by electricity but what is termed the bioplasma. In 1944, V. S. Grishchenko introduced the pos-sibility of a fourth state of matter in living beings, which he labeled biological plasma. An American physicist, I. Langmuir, also used the term in later years. In a paper, "Biological Plasma of Human and Animal Organisms," Dr. V. M. In-yushin notes:

> Until the end of the fifties no attempts were made to investi-gate the question of the presence of biological plasma in the case of living beings, in spite of the feeling that the urgent necessity for such research had constantly increased, in such diverse spheres as medical biology and psychology. Many of the biological, psychotronic and psychoenergetic, etc., prob-lems cannot be solved without a basic investigation of the bio-energetic structure of the organism and its environment, be-cause beyond any doubt every living organism is a system that is irradiating energy and creating a field around itself. [Inyu-shin, "Biological Plasma of Human and Animal Organisms," p. 50 (privately translated from the Russian)]*

---

* Translations of many of the articles by Russian scientists currently studying bioplasma may be obtained from the Southern California

Using the Kirlian photographic method, Inyushin and his associates studied in the sixties "the radiation phenomena of the biological plasma on the surface of the bodies of animals (including man)" (*ibid.*, p. 51). Radiation phenomena were found to increase with shock and excitement (in rabbits). Experiments with humans showed the radiation light to be of diverse colors and most intense at the fingertips and solar plexus. Mood influences the brightness and intensity of the radiation. In fact, it is claimed that "all the psychic as well as physical states of an animal are reflected in the total energetic state of its plasma" (*ibid.*, p. 52).

In the same article Inyushin states that the properties of the plasma "approximate to the attributes of crystalline plasma or of solid semiconductors" (*ibid.*). This point is developed further in another article. Apparently the plasma is somewhat heterogeneous, for Inyushin maintains that "all kinds of oscillations of bioplasma put together create the biological field of the organism"—with its specific biologogram (*ibid.*). Moreover, the plasma irradiates energy in "the visible spectrum or it is capable of emitting it in the form of electrons from the organism—especially when stimulating agents are present" (*ibid.*, p. 53). For instance, in an experiment, one hand was brought to within five to ten centimeters of a plant. Then

the skin of the experimenter's hand was stimulated by a needle. As a repercussion of that stimulation, an increased electrobioluminescence appeared. The corresponding increase of intensity lasted 5–10 seconds, approximately, returning afterwards to the initial level, or even to a level slightly inferior. . . . In another case (in association with the psychiatrist A. S. Romen), the subject, of unusual sensitivity, evoked a strong emotional strain in himself. In this particular case, too, it was possible to observe a deviation in the registered intensity of the electroluminescence of the subject. [*Ibid.*, p. 53]

It is significant from an orgone-theory standpoint that the Russian experimenter added, "our experiments [of this

Society for Psychical Research, Inc., 170 South Beverly Dr., Beverly Hills, Calif.

type] were unsuccessful whenever a storm was approaching"
(*ibid.*).

In a paper given by Prof. Inyushin* at a special seminar on
bioenergetics held in 1969 at Kazakhstan University, he goes
into an elaborate discussion of the nature and mechanisms of
bioplasma. As conceived by Inyushin, bioplasma basically
consists of free charged particles, which, he claims, exist in
organisms in definitely organized patterns and create uniform
energy networks. Fundamentally, they are negatively charged
particles with a small rest mass. In the following direct quote
from his 1969 article describing this plasma, one is immediately struck by similarities to Reich's conception of the orgone:

> At present one can with confidence say that almost all matter
> in the universe is found, as is known, in the plasma state . . .
> on our planet. It is sufficient to point out lightning discharges,
> the colorful phenomena of the Aurora Borealis, electric arcs.
> The sun . . . consists completely of plasma. Streams of
> plasma continuously rush from the sun into universal space.
> (It is believed that the stream of plasma fills the entire inter-
> planetary space.) That is, a stream of plasma always exists
> directed from the sun in all directions, forming a gigantic
> solar corona. . . . Under terrestrial conditions we encounter
> plasma considerably less frequently. . . . But small quantities
> of plasma in the form of air ions (Tchizhevsky 1960) are al-
> ways present in the atmosphere. Indicative of this is the faint
> light of the night sky which appears as a result of the recom-
> bination of ions. There is plasma even in a solid—conductors
> and semiconductors . . . [Inyushin, "Biological Plasma," p. 5]

This gets very close to Reich's idea of orgone being the link
between animate and inanimate matter.

Inyushin's concept of how the plasma originates is very intriguing. He says, "Ionization in a gas discharge is one of the
most widespread methods of forming a plasma. An electron
cascade is formed during a discharge" (*ibid.*). In general the
process may involve knocking an electron free of an atom.
This can be achieved through the action of an electrical field,

---

* V. Inyushin, "Bioplasma, the Fourth State of Matter, from the Point of
View of Physics." This paper was translated from the Russian in 1971.

so that there occurs "the multiplication of electrons in geometrical progression" (*ibid.*, p. 6). Examples are the ionization of oxygen through the action of ultraviolet light or photo effects—for example, the intense flash of a laser, or by X rays. Basic to this theory is the great significance of oxygen in organic processes. Oxygen has a great affinity for electrons, and ionized molecular oxygen plays a major role in biological processes. In respiration, a transfer of electrons occurs along the respiratory canal. Inyushin contends that "the electrons in the biological plasma . . . are saturated with definite portions of electromagnetic energy" (*ibid.*, p. 35). Negatively charged ionized oxygen then can be the source of free electrons—and at the same time "replenish the bioplasma balance of the organism" (*ibid.*). Thus, respiration charges the organism with energy, apart from the pure intake of oxygen and other gases—a basic Reichian affirmation. Breathing charges the bioplasma body and helps to equalize disturbed energy patterns. Breathing air with a marked surplus of negative ions has accepted therapeutic effects; it seems to help because it gets right to the bioplasma body.

Many diseases were shown to begin when the supply of bioplasma deteriorated, the Soviets say. Soviets found even spraying a wound with ionized air would greatly speed healing, as the negative ions helped restore the plasmic body to equilibrium.

"With this concept of biological plasma body, we can open new paths to understanding the growth of cancer, tumors and other forms of disease," wrote the biologists. [Ostrander and Schroeder, p. 215]

Basic to these Russian claims is the work of Nobel Prize winner Szent-Györgyi, whose researches on energy (charge) transfer are thought-provoking. (The passing of one electron to another molecule is called *charge transfer*.) He has shown in detail how some substances are donors and some function as acceptors of free electrons. He notes: "Evidence started accumulating more than thirty years ago suggesting that in certain complexes, electrons may trespass between the borderlines of two complexing molecules" (Szent-Györgyi, *Introduction to a Submolecular Biology*, p. 54). Proteins, nucleic

acids and water, apparently in some systematic interrelation-
ship, are basic to the charge-transfer process (*ibid.*, p. 93).
"The first direct experimental evidence for the participation of
free radicals in the one-electron enzymic transfer [was] outlined in the classic studies of L. Michaelis" (*ibid.*, p. 7).

In his three main works, Szent-György is concerned with
illuminating the process and problems of energy transfer in
living organisms and what this signifies. Thus, he affirms,
"what drives life is . . . a little electric current, kept up by
sunshine. All the complexities of intermediary metabolism are
but lacework around this basic fact" (*ibid.*, p. 22). While
Szent-Györgyi is very humble about his findings and talks
frankly of the many unanswered questions in biochemistry
and biology, he strongly affirms that (a) the living cell is essentially an electrical device (Szent-Györgyi, *Bioelectronics*, p.
79); (b) "charge transfer may be one of the most important,
frequent and fundamental biological reactions" (*ibid.*, p. 64);
and (c) "water is at the bottom of nature . . . It has the
lowest energy and highest ionization potential among all biological substances. . . . It is the cradle of life, the mother of
life" (*ibid.*, p. 9).

In passing, two specific claims of Szent-Györgyi deserve
mention. One is that "the cancer cell is a cell which has lost
its ability to bind [and inactivate its glyoxalase]" (*ibid.*, p.
72), which is another way of saying it has a low cohesive
force. A corollary is that it proliferates wildly. All the brakes
are off; the power to say no to growth has disappeared. This
could be interpreted to indicate that so much energy is captured in the cell—because of a surplus of undischarged orgone—that it triggers off a wild state of proliferation. The
other observation is that the tranquilizer drug chlorpromazine,
which is so successful with schizophrenia symptoms, is an
exceedingly "good electron donor capable of forming stable
charge-transfer complexes" (Szent-Györgyi, *Introduction to a
Submolecular Biology*, p. 112). From this observation, Szent-
Györgyi asks, can a lack of electrons be involved in the gene-
sis of this disease? This implies the further question, Can
bioenergetics supply, as Reich insists, a theory of and therapy
for the dread disease schizophrenia?

Using Szent-Györgyi's work as a base, Inyushin emphasizes the role of excited electrons in the entire molecular system, and also singles out structured water as a significant donor of electrons. Thus, "the presence in proteins," Inyushin claims, "of adsorbed water leads to an increase in electrical conductivity" (Inyushin, pp. 12–13). Not only can structured water be a donor of electrons and thus cause the conductivity of proteins and nucleic acids, but the water itself can produce free-charge carriers of high mobility (*ibid.*, p. 14). Inyushin notes that structured water—which seems to be present in living cells—can assist in the transfer of charges, especially protons. Their rate of migration can amount to several meters per second. Such statements recall Brunler's assertion (in Chapter Four) that protons are the source of the cosmic dielectric energy basic to living creatures, and that its movement through organisms is relatively slow. Further on, Inyushin concludes: "One can assume that water plus lipids, proteins and nucleic acids constitute, as it were, the skeleton of living matter, its most labile part consisting of free charged particles, electrons, protons and ions" (*ibid.*, p. 16). And he quotes the American biochemist Bernard Pullman, who in his *Quantum Biochemistry* has stressed the significance of molecular systems with mobile electrons, for life processes.

A number of secondary characteristics of plasma are of interest, especially as many of them seem to parallel Reich's characterization of orgone. Thus, Inyushin describes the bioplasma as naturally unstable (note that the orgone by nature pulsates) and as sensitive to magnetic fields. Also, he distinguishes two basic kinds of plasma, the dense and the rarefied. With respect to the former, he writes: "The most stabilizing action on a plasma is afforded by magnetic fields having specific configurations, for example, in the form of a helix" (*ibid.*, p. 8). In terms of the electromagnetic spectrum, he contends: "The plasma frequency can vary from microwave to ultraviolet frequencies" (*ibid.*, p. 10). "Concentrated plasma absorbs high-frequency vibrations up to ultra-high frequencies" (*ibid.*, p. 7). The higher frequencies can absorb more and therefore can attain greater density.

Some of Inyushin's conclusions are equally suggestive. Thus, he notes that "it is possible that a characteristic of the living state is the presence of a large quantity of delocalized [or moving] unpaired electrons." Quoting the work of L. A. Blumenfeld on the unusual density of unpaired electrons (almost equal to that in metals) he suggests this may imply as high an electroconductivity in living matter as in certain metals (*ibid.*, p. 15). The significance of this for the potential flow of life energy in the body is not to be ignored. Some of the final points in Inyushin's theory are summed up in the following quotation:

> Plasma in terms of the living organism, or bioplasma, is not a chaotic system, but has a very complex organization. On the one hand, plasma can be considered as a discrete system, consisting of some elementary plasma constellations [ensembles of excited electrons, protons and, possibly, other particles]; on the other hand, all bioplasma represents a unit in organism, the combination of which depends on electromagnetic and other interactions of such constellations. Biological plasma is specific for each organism, tissue and, possibly, biomolecule. Its specificity is stipulated by the spatial [3-dimensional] form of the organization, the saturability or excitability of the electromagnetic energy, concentration of charged particles, and also the velocity of their motion or drift. In the organism, drift of the biological plasma and its polarization takes place all the time. . . . The bioplasma substance is just that medium in which all kinds of electromagnetic energy which make up the basis of the biological field propagate. . . . One may assume that the combination of electromagnetic coherent oscillations and electrostatic forces in the biological plasma leads to a complex biological field with vectorial properties. It is precisely that field of the whole organism which causes the efficient and interdependent operation of the vast quantity of cells in the living organism, and their differentiation in ontogeny. [*Ibid.*, pp. 31–32]

Most fascinating in view of studies of cosmic phenomena by European biometeorologists—reported in the previous chapter—is Inyushin's other conclusion that "biological plasma causes unusual sensitive reactivity of living organisms to cosmic phenomena" (*ibid.*, p. 36). In short, the organism can respond to solar disturbances, et cetera, through the body plasma.

Another kind of data that support Reichian assumptions is Kirlian photography—of spiritual healers at work in Russia. These seem to indicate that psychic healing involves a transfer of energy from the bioplasmic body of the healer to the corresponding body of the patient. Changes in his bioplasmic body then affect the patient's physical body. A famous Russian healer named Colonel Krivorotov has been recently tested by the Kirlian method. Krivorotov, whose hands, though cool to touch, often leave a burning sensation for a day or two after a treatment, produced an unusual Kirlian photograph. Taken while he was healing, it showed that "the general over-all brightness in Krivorotov's hands decreased and in one small area of his hands a narrow focused channel of intense brilliance developed. It was almost as if the energy pouring from his hands could focus like a laser beam" (Ostrander and Schroeder, p. 219).*

From a treatise by Dr. Victor Adamenko, published by Novosti Press Agency, under the heading of Bioenergotherapy, there is some further interesting information about Colonel Krivorotov's healing:

If the patient is healthy, he feels at 5–10 centimeter distance a "subjective" heat from A. Krivorotov's hands pleasantly spreading throughout his body. The sensation of heat corresponds to a temperature of 45–50°C and cannot be measured objectively. Some people experience the feeling of light prickling, easiness, intoxication . . . As A. Krivorotov slowly passes his hands at some distance along the patient's body, there arises in the patient approximately at the site of the sick

---

* As this book was going to press, I received information of a symposium on psychic phenomena in healing, held October 5–8, 1972, in Los Angeles, and sponsored jointly by the U.C.L.A. Extension Department of Health Sciences and School of Medicine and the American Academy of Parapsychology and Medicine. Speakers with reputable medical and pure-science backgrounds discussed psychic healing, acupuncture, hypnosis and biofeedback, and many saw these as exemplifying a common energy system. Dr. Thelma Moss, a researcher at U.C.L.A. Health Sciences Department now involved in duplicating Russian Kirlian photography, showed Kirlian-type photographs of psychic healers *before* and *after* doing healing. These showed that the aura, or electric corona, around the healer shrank after a laying-on of hands, while that of his patient expanded.

organ a strong subjective sense of heat, at times almost unbearable. A. Krivorotov also feels at this place an intensification
of heat in his hand . . . A. Krivorotov's hand remains at the
sick organ until the sensation of heat becomes annoying. It is
a signal for the session to be terminated . . . Perception of
"heat" in different people is different and is, apparently, related to whether the disease is more or less serious. . . .
Sometimes the subjective sensation of heat at the site of the
sick organ persists in the patient for two days. However, in
the process of treatment there gradually sets in an adaptation
to A. Krivorotov's field . . . Remote "heat" is not perceived
by everyone . . . An official test of the treatment method . . .
was conducted in a clinic in 1956 on the decision of the
Learned Council of the Georgian Republic's ministry of
health. The conclusive evidence obtained by a commission presided over by Academician Pyotr Kavtaradze was positive.
A. Krivorotov deals primarily in nervous diseases . . . Rather
quickly cured is poliomyelitis, but hypertension, bronchial
asthma, polyarthritis yield very slowly to treatment, the effect
being at times completely absent. Similarly to A. Krivorotov,
the method of bioenergotherapy is possessed by his sons—
Vladimir, a physician; and Victor, a mechanical engineer.

An analysis of A. Krivorotov's work has revealed that par-
ticipating in the process is a strong electrostatic field. Occasionally slight discharges take place between A. Krivorotov's
fingers and the patient's body. Subjective sensations of heat
cannot be produced by a low-power source of electrostatic field
as is A. Krivorotov. . . . Therefore, it is obviously a question
of a reflectory effect of an electrostatic field from A. Krivorotov's hands on the patient's skin receptors, whilst a prolonged
feeling of "heat" persists in the patient's memory. . . . High-
frequency (Kirlian) photos of A. Krivorotov's hands in a high-
frequency discharge field [provide] experimental evidence in
support of the electrostatic field . . . generated by A. Krivorotov. . . . The sole conclusion to be drawn from this is that
either A. Krivorotov's field is qualitatively different from a
technical field, or it merely accompanies the effect of some
other agent. In any case, there exists a correlation between
A. Krivorotov's electrostatic field and the subjective sensation
of heat, produced in the patient. [Adamenko, *Electrodynamics*

*of Living Systems* (Moscow: Novosti Press Agency, n.d.), pp. 115–16]

Perhaps the most amazing energy development in the Soviet-bloc nations is their discovery of the way in which specific shapes, like a pyramid or a sphere, somehow convey a kind of force. This recalls the yoga belief, discussed in Chapter Six, in five kinds of ethers, each possessing its own energy potential. The Czechs discovered, for instance, that razor blades left inside a small pyramid-shaped box retained their sharpness much beyond the normal usage. This device was patented in 1959 and is selling well. The origin of much of this new interest is a book by a French engineer, L. Turenne, *Waves from Forms:* "Various forms, Turenne asserts— such as spheres, pyramids, semi-spheres, squares—act as different types of *resonators* for the energy of the cosmos, the sun, and the energy all around us. Just as the special shape of a violin gives tone and quality to a bow touching a string, the special shape of a pyramid apparently is a resonant cavity for the 'live' crystals of a razor blade" (Ostrander and Schroeder, p. 360). This raises probing questions regarding the Egyptian pyramids. Did they embody energy functions?

Czech scientists have perfected an apparatus, called a psychotronic generator, which is supposed to draw energy from a person, accumulate it and use it. They claim such generators can do some of the things that psychics do, such as psychokinesis. They claim their apparatus accumulates bio-energy. The inventor, a design director named Pavlita, supposedly created his energy accumulator through delving into ancient documents. An early form of his device, a lightly sealed metal box with a tiny psychotronic generator inside, was first tested by an electrophysiologist and numerous physicists at Hradec Králové University, east of Prague. A few years ago the Central Committee of the Czech Communist Party approved research on the apparatus. One of its forms is used to radiate bean seeds so they will grow bigger. In this experiment, the psychotronic generator, a studded metal square supporting a coiled, borelike neck, is directed at one of the pans of seed. Some days later, the plants from these seeds have grown to twice the size of the untreated controls. Other generators

purified (dye-filled) polluted water, and gave results in tele-
kinesis, telepathy and clairvoyance. Some Russian scientists
have shown interest:

> Dr. Genady Sergeyev, the Leningrad neurophysiologist, com-
> mented . . . "The Pavlita work shows it is possible *to trans-*
> *fer energy from living bodies to nonliving matter.* The most
> important influence of this energy is on water. In fact, we
> use this very principle in the development of the detectors that
> examine the fields around Mrs. Mikhailova during PK." Both
> Sergeyev and Rejdak talked about electronic bioplasma in the
> human body as the basis of PK and this psychotronic energy.
> [*Ibid.*, p. 374]

Apparently the secret to these generators lies in their form.
Russian scientist Dr. Rejdak claims they are an extension of
von Reichenbach's idea of an odoscope—an apparatus to col-
lect odic force—but greatly modernized. They are of two
types, those that accumulate energy from living things and
those that extract cosmic energy from the environment. The
pyramid is an example of the latter. The former take energy
from the human force field, sometimes by a process of staring: "Many [generators] have a certain staring pattern carved
into them to help concentration and conduction of energy"
(*ibid.*, p. 376). New types of generators work without any
staring.

EEG tests on subjects "charging" a generator show changes
in brain patterns. According to the Czechs, this psychotronic
energy has many similarities with electromagnetic energy, but
is much more subtle. Pavlita's generators do many things, including the turning of motors. If focused on people they seem
to speed recovery from wounds and various illnesses. The
more one reads about them, the more one sees parallels with
Reich's claims for orgone energy, the accumulator and the
orgone motor. Have the Czechs stumbled upon another way
of tapping orgone, or have they found an entirely different
energy? Pavlita even stresses the dangerous aspects of an over-
dose, just as Reich warned about too much orgone. Since
copper and steel are used in Pavlita's generators and sometimes steel is joined to a nonmagnetic substance like wood or
plaster, there are even similarities in construction between

them and the orgone accumulator's components. Of course, they could be accidental. In lieu of actually seeing these generators and discovering the basic principle of their construction, one can only say that their so-called powers add some plausibility to Reich's claims for the orgone.*

* Reich in his later days was constantly worried that the Communists might develop and exploit his ideas, and some Russians did order his books in the early fifties.

# PART
# VI

*Chapter Sixteen*

TESTING ORGONE BLANKETS

This chapter describes specifically my personal experience with tests on orgone-energy blankets. I began these in the mid-fifties, when there were few detailed studies on the orgone's therapeutic and energizing effects. Even today, apart from articles by medical orgonomists, one has to search far and wide to discover independent experiments on the orgone's effects on people. While my tests and experiences with orgone blankets were not, in a strict sense, scientific, the range of effects and the time devoted to these efforts have undoubtedly helped me reach conclusions about the orgone theory. These will be discussed in the next chapter.

I did try to conduct controlled tests with cancer-infected mice, but a doctor who promised his cooperation changed his mind after the mice had been purchased and the physical arrangement well prepared. He was afraid to get involved. His professional career might be damaged. Later, when I inquired of friendly individuals on the margin of in-depth medical research, I was informed that it would take scores or hundreds of simple controlled experiments with animals to impress the medical establishment, which meant a huge budget and a staff of researchers. This news, which I took to be accurate, totally dismayed me. Being a sociologist, I had no chance of getting any sizable grant for such research. So I carried on simple

ORGONE, REICH AND EROS

checks on orgone blankets with family members and friends.

I realize that the following account will convince no hard-headed scientist. Suggestion or coincidence will be their explanation. Others may find some fascinating parallels and similarities to the claims of Reichians. As for me, my testing has made it somewhat easier to grapple with the two questions, Is there an orgone energy? and if so, How does it actually work?

Let me now report, as carefully and objectively as possible, the results obtained over the past sixteen years with various types of orgone blankets. While these results are skimpy and inconclusive, to me and to others they seem to have some suggestive value. First, let me describe the most objective and interesting group test that I have had a chance to make. This was carried out at a seminar on spiritual healing, in June of 1956, sponsored by the Laymen's Movement for a Christian World. The weekend seminar, one of a series on unorthodox healing held at Wainwright House, Rye, New York, in the mid-fifties, drew about forty-five professionals in theology and the healing professions and included the novelist Aldous Huxley. I was permitted to make a controlled test during one of the lecture periods.

For this experiment I made two three-layer orgone blankets (layered with wool and steel wool), and two other dummy blankets of the same woolen cloth and looking the same to an observer, but layered inside with newspaper instead of steel wool. Although the blankets were a little different in feel and in weight, participants only saw the four and had no chance of making comparisons by touch. Allowed by the chairman of the session to make a brief introductory announcement, I simply said that I was passing around a special kind of blanket which those who so wanted could rest on their laps while listening to the lecture, and that after keeping it there for *at most* a half hour, should pass it on to someone else nearby. I said that two blankets were dummies and two the real thing, and told them I'd be asking for a report on their reactions at the end of the lecture. (There were whispered discussions about the blankets during the hour of the lecture.) I said nothing about them being orgone or energy blankets and only two or three members of the group knew this from my previous conversations with them (and these did not use them); nor was anything said as to the kind of sensation that might be expected, nor, of course, was any clue given as to how to distinguish the dummy from the real ones. All this was done to rule out, as far as possible, the suggestion factor. (I might add that very few people there even knew of Wilhelm Reich and almost none were aware of his orgone theory.) The blankets were passed around in a random order and I sat with my back to most of the audience, so they would receive no clue from my facial expression.

At the end of the lecture, I was allowed time to ask those who had had blankets on their laps to report their reactions. In every case (6–7) those who had taken the *dummies* said that they felt nothing. The reactions of the others, about ten in number, were very interesting. Several said they felt a strong sense of well-being; others felt some warmth; Mr. Huxley, for instance, said it made him feel quite relaxed, and later on asked me for one. In a letter he wrote subsequently, he referred to this experience indirectly as follows: "Just returned from a Conference on 'Unorthodox Healing' attended by a group of psychologists, psychiatrists, neurologists, theologians and two sensitives . . . Some evidence tending to confirm von Reichenbach's hypothesis (revived in our day by Wilhelm Reich) was mentioned" (*The Letters of A. Huxley*, pp. 749–50). One or two others said that they felt some warmth. Startling and suggestive reactions came from the sensitives present: Mr. Ambrose Worrall, of Baltimore, the healer, his wife, and Miss Farelly, a radionic* machine operator and former teacher of biology. All reported, with absolutely no opportunity for mutual collaborative checks on their reactions, what I had personally observed—namely, that they could not keep the blanket on their laps more than five or ten minutes, because they were beginning to get nauseated

---

* Radionics, like radiesthesia is an emerging field of ESP, which, while lacking so far orthodox scientific support, is a discipline concerned with the perception and understanding of radio-type emanations which carry information and can be picked up by sensitives who operate an Abrams- or Drown-type machine. Its main function seems to be to help the operator focus in on otherwise unknown microwaves which carry data on an individual's health or emotions.

or uneasy. (None of the sensitives, unfortunately, received a dummy.) At the end of the lecture, Mr. Worrall, a very healthy, ruddy-faced man, was still white "under the gills" and asked leave to get a glass of water. All three told of feeling a strong energy beginning in the solar-plexus region, and rising up slowly until they began to feel sickish. I was sitting next to Mrs. Worrall during this lecture, and she had hardly placed the blanket on her lap, when she urgently whispered to me, "What have you got here? I feel some kind of nature force here!" There followed other whispered comments describing her feeling an energy, "something like electricity," rising up; in less than ten minutes she quickly handed the blanket to her husband. Let me note that neither she nor her husband nor Miss Farelly had *any knowledge of Reich or his orgone theories* prior to this test. They had not discussed orgone energy with me nor had they known me at all previous to this conference. To them I was a stranger from "up there in Canada."

The reactions expressed by other members of the conference germinated some interest in the blanket, and I had requests for blankets from four people, including a leading New York theologian suffering from chronic backaches, and Mrs. Margueritte Bro, a well-known American author and editor with Harper & Row, publishing company. So much for that test.

During the fifties I gave the blanket to other psychics with much the same reaction. For instance, very shortly after meeting Mrs. Florence Baxter, now deceased,* and before telling her anything about Reich or indicating anything of the kind of reaction she might expect, I asked her to put a similar three-layer blanket on her lap and report her feelings. I wrote them down verbatim as they came. Here is how they began. The blanket was on her lap less than half a minute when she began.

> It's got an electric charge! It's burning hot! It gives me a wonderful feeling; now I'm starting to see something; I'm taken to a green field, people are dancing, are very happy; I see a beautiful tree, with a huge trunk; I think I see fairies dancing

---

* She and her husband co-authored a book, *Many Roads, Many Mansions.*

around and a lovely stream. . . . I feel lightness, something
magnetic or electrical up my arms; [she had her hands on the
blanket which was resting on her lap]; I get a sensation like
before a thunderstorm—like electricity going through me. I
feel a great well-being and lightness of body, as if I had no
body. I have a weak spot in my neck [which was at one time
quite arthritic], and now it is all tingling and my fingers are
tingling too! I think that this pad if placed on arthritic people
would give a healing effect. . . . It can also draw poison out
of the body—that is, infections. Why, this is an amazing
thing! I don't know how or why it's been charged, but this
thing is alive! There is something here that lives. . . . I feel
as though I've had an electric magnetic treatment. . . . I was
tired, but I'm not tired any more.

With a few inconsequential omissions these are her exact
words as they were uttered over a period of about ten minutes, with no prompting on my part.

Early in January, 1957, I met James Wilkie, later to be
recognized as one of Canada's most interesting clairvoyant-
healers. We met through the graces of a mutual friend, and
shortly after being introduced I asked him to put the orgone
blanket on his lap; his response as written down at the time
was as follows: "I think of an X ray; this blanket in its usual
form is bigger and not so thick as now." (I had given it to him
doubled up, and in addition the steel wool had got bunched
up at one end, so this observation was quite correct.) "This has
healing activity, which it gives to people; the sick draw physical energy from it. There is a linkage with higher forces; it
gives beginning healing for suffering ones. Spiritual healing
follows this path. I feel a physical healing energy here; almost
a magnetic energy . . . Keep studying this and its [causal]
conditions. There are changes yet that will increase the healing
powers. The healing garment [he called the blanket a garment]
should be charged from your hands, in your own home, before
being taken to a suffering one. This is only a beginning; there
is much more to be done. I feel it is still in the experimental
stage. There are rays here which open one up to healing. It
seems to help mental cases as well." All this, with several
pauses and some references to people helping on the other

side, including Dr. Kilner, took but two or three minutes. Wilkie, of course had no knowledge of Reich or the orgone.

Two other persons that I met in the fifties may be quoted. One is a former neighbor, a young Scotswoman who, I learned subsequently, has a small gift for clairvoyance. Within a few seconds after placing the blanket on her lap she said, "There is some kind of electric charge here. I feel it very easily." She then mentioned its warming qualities. Subsequently, through a daily application of about half an hour, she made a very fast recovery from double pneumonia and returned to work much sooner than she or her doctor had expected.

The other was a Mrs. Adams who gave aura readings professionally in Toronto in the fifties. Again, with no introduction or information about Reich, or about what to expect, I put the blanket on *my own lap first*, and asked her to tell me what she saw. Again, I wrote down at once, while she was speaking, just what she said. "There is a musical vibration there," she began. Then, "It makes a circuit with your body; it has golden-colored flickers at the edges; it is a gray color and has a low vibration." Prodded to tell more, she said, "It gives a pleasant and warming sensation; it gives off some kind of spirit. One should meditate while one has it on." All this was said in the course of two or three minutes. I asked her to put it on herself, giving no indication of how true her comments were and she said at once, "I feel a warmth and a stimulation. It tends to take the consciousness up. It has a good effect on the breathing. My head would soon start to swim if I kept it on very long. People with asthma and bronchitis should find it very helpful." With this, and inside of three minutes, she urgently returned it to me. It was evident that she had had enough! Any more would have upset her.

These experiences and many others suggest that people with strong psychic tendencies are the most sensitive to the orgone blanket. This may be part of the explanation of the marked reaction which it brought from my second-eldest daughter, Gwynneth, seventeen, who, according to Mrs. Baxter—and our own observations for what they are worth—is perhaps a little psychic. In 1956, after some weeks of using the blanket on myself and my wife, I decided to see if it could do anything for this daughter, who was then nine months old and certainly not a happy child, especially in comparison with her older sister. From her birth, Gwyn had been considerably tensed up, occasionally colicky in the first six weeks, and now at nine months she still cried a lot. At the same time she failed to exhibit anything like the physical mobility we had observed in our first girl at the same age. (I kept a simple daily diary on the two girls' physical and emotional development for the first twelve months, and comparisons were therefore easily made.) For instance, at nine and a half months Gwynneth did little crawling and had yet to get up and walk around the outside of the playpen, actions which my first daughter, Jocelyn, had performed at six months.

We decided to see what kind of energizing powers the blanket had, since Reich claimed orgone could have amazing effects on children. Each day for about a half hour we put it on top of her regular baby blanket as she slept in the afternoon or early evening. Within a week, she was a lot happier, and also had become much more mobile; within two weeks she was a much different girl and was walking around the playpen every day with great ease and delight. The most interesting result of this treatment, however, was that after a few days of this orgone irradiation, we were suddenly faced with her awakening almost every night, about one o'clock or one-thirty, overflowing with energy and smiles and *happiness.* Now, although this waking pattern may seem nothing unusual for a young baby, Gwyn had not done this for the previous six or seven months. But most interesting to us was that on these occasions she simply would not sleep; we would put her back below the bed-covers and tuck her in most carefully, but in a few minutes she was up at the cribside again, jumping around and wanting to play. It would usually take my wife and me an hour or so to get her to settle down and go back to sleep. All this between one and two in the morning.

We got increasingly impatient as this went on for night after night and finally decided it had to stop; we needed our sleep! We searched our minds for an explanation of this unusual behavior and finally it occurred to me, it *might* be the orgone blanket. So we made a test; we stopped the orgone treatments; immediately she stopped the waking-up business

*and never went back to it afterward.* (All together she had
had the *daily* irradiation for about two weeks.) Later on, I
gave her treatments one day here and another day there, but
not in daily sequence, and there was no repetition of the
nighttime wakening. Within a couple of months, she was a
different girl, just as happy and outgoing and mobile as her
sister. My wife, who was most skeptical in those days about
the blanket, grudgingly had to admit it had really helped
Gwyn. At two and a half years, Gwyn seemed even more outgoing and happy than her sister. To me, since there was no
chance of verbal or symbolic suggestion in this instance, this
specific experience was an interesting demonstration of the
energizing and mobilizing powers of the blanket—at least for
one person.

After this experience with Gwyn, we used the blanket
around our house quite frequently. In the next ten years, with
all our four children and usually with adults, we found it
stops colds if used immediately after the first symptoms appear. This has been our experience on dozens of occasions.
The best place to put it, for colds, is over the lap and feet.
After a cold or cough has got a good start, however, the
blanket not only fails to help, but seems to aggravate the
condition, at least temporarily. We also used the blanket to
reduce backaches, and my wife, who has these occasionally,
has secured quick relief numerous times; usually it sends her
to sleep in a hurry and when she wakes the ache is gone. Occasionally, when they were very young, the blanket (over the
stomach) was used to stop the children's crying, and almost
invariably it worked.

After some months of fairly regular use our older girl,
Jocelyn, developed a liking for the blanket, and occasionally—
when apparently quite tired—would ask for it or get it herself.
For several years my wife had little faith in it; she was
skeptical of Reich and all psychoanalysis, and so I tried a
special test on her. Since, according to Reich, during electric
storms an orgone accumulator or blanket will draw in a heavy
charge quickly, a charge so large it may induce nausea (normally the result of too long an irradiation), I waited till an
electric storm came and then asked my wife to put a blanket
on. She kept it on for about a half hour while reading, paying

no attention to it. Of course, I had not told her what to expect. Shortly afterward she began to feel a bit woozy, and she had a lot of wind on her stomach. The next day, she still had considerable wind on her stomach and belched up air for several hours although this is *something that she never does normally*.

Enough of my family. I shall skip my own reactions, because they could be due to suggestion. Suffice it to say that my reactions to the blanket are less pronounced now than they were originally, ten or eleven years back, but it still gives me an energy pickup in fifteen to twenty minutes.

Over the past few years I have made perhaps sixty small three-layer and three six-to-eight-layer blankets and loaned or given them to friends. Here, very briefly, are some of the things reported to me. Gord L., a writer, who has had a long-standing health problem, reported that he seemed to have a lot more energy than formerly. Lying down with the blanket for half to one hour per day allowed him to keep going at night two or three hours extra without tiring. This went on for a long time. The unresolved question was, Did the extra energy come from the blanket or from the lying down and resting that went along with an irradiation? Sylvia W., a social worker, who kept one for a year, reported that on a number of occasions it helped to relieve recurring backaches. My mother, who had a blanket for years, reported that it seemed to give her a little extra energy and that it helped her to get to sleep at nights. (She has been troubled with insomnia for years.) Delchen D., a woman in her sixties, with extraordinary liveliness for her age, but unable to get to sleep for an hour or so after retiring, also reported that putting a three-layer blanket over the bedcovers got rid of her insomnia.

Some years ago, when holidaying in Windsor with relatives, we were constantly bothered by a next-door neighbor's baby, who cried every day, literally by the hour. In desperation, we finally loaned its mother a blanket, and for the duration of our stay, about a month, we heard practically no crying. My stepbrother, then eleven years of age, became bothered with strep throat, so we got him to put on the blanket two days running for about a half hour. In two days the infection cleared up and his mother, originally very skeptical, was quite impressed.

A friend of one of our Toronto neighbors was suffering from polio. He had a crippled arm and leg, and couldn't get around. After one month of daily irradiation with a six-layer blanket, he was almost fully recovered and able to start back to work. The speed of his recovery surprised both doctors and family. A young woman friend wanted to try out my big eight-layer wool blanket (5 by 2 feet) and after a half hour, in which she began to breathe very deeply and felt twitchings, developed a sharp pain in her left shoulder. It got worse the next day and she had to go to a doctor. (She had obviously had too much, and it had focused on an area of armoring.) The mother of my secretary, a woman in her late fifties, developed sciatica one winter and I lent her a three-layer blanket. She reported that while placed over the area it took away the pain, but the discomfort usually returned about an hour or so afterward. Later, the doctor diagnosed the complaint as arthritis; she continued using the blanket daily and within a few weeks the whole complaint had apparently cleared up. She asked me to sell her the blanket, as she wanted to have it around permanently.

Mrs. A., a lady of sixty-five and in feeble health, broke a leg and experienced a lot of pain while it was healing. She began to use the blanket daily, and when I went to see her some weeks after, with no prompting or questioning she said, "That blanket was a lot of help; it took a lot of the pain away." Her cook, Mrs. W., was suffering for years from bad nerves and general tiredness. Now she uses the blanket from time to time and reports that it helps her to relax and makes her feel a bit better.

John W., a very healthy and sturdily built young man of around twenty, asked to be given a test, and he put a six-layer blanket over his lap and chest, while sitting on a three-layer one. We chatted and kept checking on his feelings; after over an hour and a half, he reported feeling rather relaxed. This suggests that some people feel little or nothing and corresponds to Reich's own observation. A woman friend sat through a meeting, lasting two and a half hours, with the blanket on her lap. All she felt, when I asked her at the end, was that it was hot.

It is clear from the above that reactions are varied and in

most cases mild. In explanation of the latter, it ought to be noted that while Reich usually operated with many-layered accumulators I normally use only a small, three-layer blanket (2 by 2½ feet). According to Reich's theory, it would have little power. On the other hand, since he advises using more than three-layer accumulators only with strict medical supervision, I believe it is wise to keep to the weaker three-layer type. Whether much more can be accomplished with eight- or ten-layer blankets, at this point, I am unable to say. The last person to use my six-layer, 5-foot-long blanket was Ernie S., who had multiple sclerosis. He reported at once that it made him feel relaxed and comfortable; later, he began to talk of some energy or charge flowing from his hands when he brought them together. This may or may not have been due to the energizing of his system. After several weeks' use, it apparently did nothing for Ernie's basic condition.

Just prior to writing this book I made a three-week trip to the United States and took with me an eight-layered blanket, determined to test it on every possible opportunity. The first major chance came when I was invited to another weekend at Wainwright House, in Rye, New York, where the Worralls were lecturing on "Psychic Gifts." One of my hopes was to retest them with the stronger blanket, but they had not forgotten their reactions to the small, three-layered one—even though fourteen years had elapsed since that experience— and in spite of my requests politely refused to cooperate. I won't say that they were frightened to try it out, but they certainly expressed no eagerness to even put their hands on it for as little as five minutes. (And yet they are not critical or contemptuous of Reich's ideas.)

While waiting for a final answer from the Worralls, I thought I would just pass it around casually before their Saturday afternoon lecture. I had arrived there only an hour or so before, a stranger from Canada, and knew none of the forty conferees, nor did they know anything about orgone. So, while we were waiting for the lecture to begin, I asked a young man active in the Spiritual Frontiers Fellowship (he said he was a healer) to put it on his lap and report his feelings. Within a couple of minutes he exclaimed, "I have a light feeling in the head—a slight detachment." In a few seconds,

"Now I feel a lightness over all my body." Then after a half minute, "I feel a tingling in my hand" (one hand was resting on the blanket). "There is some action going on there—some kind of drawing." After about eight minutes he said, "I feel very heavy," and he began a series of big yawns. Questioned, he revealed he had not had the usual amount of sleep the previous night. Apparently, the blanket had made him relaxed and quite drowsy.

A female friend of the young man became curious, and I said, "Try it on; the effects are not always the same." She claimed some slight psychic gifts and was obviously a very sensitive and intelligent woman. In a minute or two, she said, "I feel a strong force here—it has healing power—I feel it in my legs." As she got interested and had it on a little longer, she added, "The healing force has gone down to the ankles." (She had the blanket over her lap, and part of it was hanging down toward her feet.) After a while she added, "It's like being in meditation. My legs feel totally relaxed." After about ten minutes, she too had had enough. "I feel drained," she said.

Other persons who were handy were now interested. A young woman of thirty named Jennifer asked to have a try. In a little while she said, "I feel an intensification of energy in my body. I feel some tingling in the upper legs." Surprised at this, I asked about her background, and I discovered that she had recently had a few hours of bioenergetics with Dr. John Pierrakos and had begun to feel energy flows in her body.

Others who gave the blanket quick trials and whose names I did not catch, made the following comments, some orally, some in written notes passed to me during the lecture: "I get a tingling in one hand [the hand on the blanket], and I felt some energy coming in [after five minutes]." A woman tried it over her back and felt nothing after at least fifteen minutes. "Felt considerable warmth over my entire body at first. I felt comfortable, but after five or ten minutes the thing became cold and heavy, and my legs got restless," said another. "I felt chills and prickles, then just heat!" "I felt warm, terribly relaxed, and quite sleepy."

The experience with a doctor's wife, which follows, has

special implications, particularly in view of a subsequent test with Mrs. Ann Armstrong, a clairvoyant living in Sacramento, California. The doctor's wife (Mrs. D.) had noticed the blanket and people trying it, and after a while she said, "Can I try it?" I didn't know her at all and answered, "Why, yes, of course!" My back was to her, and I was conversing with other conference participants when I heard someone crying and sobbing behind me. I turned around and there was Mrs. D. crying and urgently gasping for breath. I went up to her and said, "What's the matter?" She had had the blanket on her lap no more than two minutes, and I was completely puzzled. As clearly as I can remember she said, "My heart is pounding terribly" (I felt this myself) "and I'm frightened. I just put the blanket on and after a minute I felt I couldn't put it off, even though I wanted to. It seemed like a magnet, glued to my lap." (By this time she had shoved the thing onto the floor.) "I feel like I'm going to die or something," she added between sobs. I was at a loss to explain this. She looked like an ordinary housewife in her late thirties; she was not a psychic or healer, and she didn't appear to be a hysteric.

Somebody said, "Let's get her to Ambrose Worrall," and so she was assisted upstairs to the room where he was resting. (His wife was giving the presentation that afternoon.) When I saw Mrs. D. later in the afternoon, she said that Mr. Worrall had claimed that something had got twisted in her back and he had put it straight; this was apparently the source of her discomfort. (In his healing, Worrall puts hands on the body, and from his experience of the last forty years he has gained some diagnostic understanding.) Had the orgone shot into her so strongly that it caused a slight dislocation in her back? The other thing Worrall told her was that her aura was exceptionally large, extending three feet out from the body. An implication here was that she might have unusual psychic powers.

During the balance of the conference and subsequently, I discovered several pertinent facts about Mrs. D. The mother of five children, she is a person with unusual energy; she regularly meditates from five to six in the morning and then rises to get at the day's activities. She is a leader of a small group which does yoga every week. She has had numerous experiences of spontaneous precognition, and has done some automatic writing in recent years, but has never considered herself a medium or gone to spiritistic meetings. In short, she may well have some strong psychic gift, or an unusual force field around the body.

In California, about ten days later, I heard from a graduate student of Berkeley about a middle-aged clairvoyant, Mrs. Ann Armstrong, and resolved to ask her to try out the blanket. We met just before she was to address a home meeting about her psychic readings, and I secured her consent to try an "experiment" with the blanket at the conclusion of her speech. Mrs. Armstrong, who only became aware of any psychic abilities a few years ago, is a small, slight, Spanish-looking woman in her late forties. (She was one of a number of respected Californian psychics scheduled to address a two-day course on parapsychology in February, 1971, at the University of Southern California at Davis, under the sponsorship of Professor C. Tart.) Before the meeting I told Mrs. Armstrong, who had never heard of Reich or orgone, just enough about the blanket and its possible energy capacities to arouse her curiosity. Nothing more!

At about 10:15 P.M. Mrs. Armstrong had finished her speech and questions and indicated she was ready for the experiment. I then produced the blanket and took it to where she sat, while the group of about twenty persons watched. I had given her no instructions on how to use it. She seemed to want to put it between her shoulder blades and so she put it on the floor and lay down on top of it. She said, "It should come down to the base of the spine," and stretched out on it, so that it covered both shoulders and the upper pelvic area. I took exact notes of everything she said and what I observed her doing.

In a matter of seconds, she began to breathe deeply. In ten seconds, she gave little grunts, as the breaths became extraordinarily deep. At fifteen seconds, she said, "I'm getting very hot" and in a few more seconds, she gave some little gasps and cries—as if there was some slight pain. "There is more flow in the left side of the spine than the right side," she exclaimed. More deep breathing followed. Then she moved herself up further on the blanket.

"There is an energy field in the blanket and the energy field in the body is trying to equalize with it," she continued. "It's pushing me away. I'm building up a tolerance in my body so I can take whatever is there." By this time, she had lain on it for about two minutes. Now, I observed her toes flexing and bending in and out. (She had taken off her shoes at the beginning of the test.) This is the kind of wiggly-toe phenomenon that Bob Hope humorously associates with a long and delicious kiss, or what kittens do when they are nursing.

"Now I feel there is an equal flow on both sides of the spine—and I'm getting relaxed," she went on. Then after a little while, "Some part of me is separated. I have a feeling that part of me is floating over my body. And I feel good about it." (This doubtless refers to the sensation of lightness that the healer at Wainwright House had noted.) "Now my chest muscles are beginning to loosen up." Then after a few seconds, "I can experience my auric field and I can feed off that energy in the blanket."

Her next comments were equally interesting: "I feel a shaft of energy going into the top of my head, going down the spine and out my feet." Then after a few moments: "I now feel through this I am connected with the earth; I like it! I'm getting in touch with earth currents and I like it!" Up till now, about three minutes had passed. Then, "At the base of the spine, I feel I want to push into the floor and into the blanket, and get into it!" And she made pushing down motions with her pelvic area. Now, another surprising comment, confirming basic Reichian theory (from someone who had never heard of him): "Before I didn't feel I could take it there . . ." (meaning in the pelvic area).

Deep breathing now became very visible. At the four-minute mark she was totally involved in very deep and quite fast breathing. She said nothing for about a minute, but seemed absorbed in the deep breathing which was extraordinarily heavy and fast. Then after a minute and a half, approximately, the breathing slowed down and became less deep; she got up and moved away from the blanket, sitting down deliberately away from the blanket. She said, "I feel like that is all I can take from it. . . . I feel good; it gives me a light and exhilarating feeling. Why, it's like the negative ion charger. It also reminds me of the Faraday cage." (She had been previously tested in a Faraday cage.)

As she seemed quite ready to talk, I asked her some questions. "What is the source of the energy, do you think?" I asked. "It's partly from the material in the blanket, and partly from the air," she replied. "It works like a battery or condenser. Different locations [geographical] can give different reactions . . . It is responsive to different weather conditions. It will give a different response in the sunshine. The sun and dampness would probably provide different reactions." All of these remarks are, clearly, confirmations of Reich's theory —from someone who had never heard of him!

My next question was, "How can this energy help the average person? How does it work?" She answered, "At first, it came in on the left side and built up a kind of overcharge, then began to flow through the right side, and then I began to relax. If the body is out of harmony, there can be trouble." (This is what may have happened to Mrs. D.) "The energy has healing qualities; it is beneficial. I could benefit from it myself; it acts to restore harmony to all the cells of the body. If an individual is too much out of balance, it might be harmful. At the end, I felt that I had so much energy that I was slipping out of my body. It was like taking some psychedelic drug . . . At first, I had to keep it [the blanket] from the base of the spine. . . . The first charge went down to my legs; then I felt, a bit later, that I could put it [the blanket] down further." (This was when she moved up on the blanket so that it reached farther down toward the base of the spine area.)

Then I asked, "What do you mean by slipping out of your body?" "I mean I would go into a catatonic state . . ." Finally, after some minutes of this questioning, I asked, "Just how do you feel now?" and her reply was simple: "I feel warm; my arms feel kind of prickly; and my back feels warm . . ." And with that our discussion came to an end. Without knowing it she had confirmed in a few minutes much of the basic theory as to how the orgone energy works—and also thrown light on why Mrs. D. had put out her back in the first minute of irradiation.

Shortly after returning from the United States trip, I was

asked to give a lecture to a York University humanities class on Reich, and decided to try another test. At the beginning of the lecture, I passed out two orgone blankets (calling them pads), refusing to explain what they were. I said, "Put it on your lap for a few minutes, see what reaction you get and then pass it to a neighbor in the class." In the lecture I dealt with many aspects of Reich's life and work but deliberately refused to say how an orgone blanket is made, what the orgone "feels" like or what feelings it may arouse. (Obviously the suggestion of an energy was there, but not how one perceives it.) After finishing, I asked those who had had on a "pad" to report—some seven or eight had felt nothing, and three described mild prickling and some warmth. One girl, who did considerable artistic dancing, also felt an energy flow in the legs and thigh area.

While writing this book, I moved into residence at the Glendon campus of York University and continued some testing. Professor Don Carveth in sociology, giving a course on youth, had several students who were interested in Reich. One was a girl of nineteen who claimed to be able to see auras around people's heads from time to time. I loaned him a blanket. Several students came to his office and put it on. They felt prickling and some moving energy around the upper leg. The aura seer kept it on less than five minutes and said it was making her feel dizzy and shoved it off.

Finally, I asked a girl to whom I had been only introduced briefly if she would try an experiment. She seemed more than ordinarily vital, and I was curious to see her reaction. She brought two other students to my room and I recorded their reactions. None of these students knew me or had read Reich. The first student, Sharon K., told me she had taken just an hour or so before a big shot of tranquilizer, on her doctor's orders. She put the blanket on for twelve minutes all together. She felt only a tingling then, but shortly afterward reported: "It's like being hypnotized"—and she dozed off in a chair. Her girl friend, Cherry B., made the following comments in the course of ten minutes. "Got a prickling in my fingertips and right down to my feet . . . I'm feeling hot . . . I've got a buzzing in my ears . . . and now a pulsing at the back of my neck. I'm very aware of my body. This would be good for meditating. I'm taking deep breaths—I'm getting charged up! There is a tingling in my fingers" (she had her hands resting on the blanket, which was on her lap). "I'm beginning to feel relaxed. I've got into a definite rhythm . . ." After she had given the blanket to Ike K., she said, "I feel like something has been put into my system."

Ike, a big, energetic young man of about twenty-one, made the following comments in the course of fifteen minutes. With one arm resting on the blanket on his lap, he said, "I feel a current flowing up my arm. It's very relaxing." After a couple of minutes more, "It has great warmth. I feel a prickling. I'm breathing more slowly and deeply. I have a definite feeling of energy coming through the blanket . . . the prickling is getting more intense in my hand [the one resting on the blanket] . . . Now I'm getting a quivering feeling in that hand and arm . . . This gives a sensual feeling . . . Now I'm developing some tension in the back of the neck . . . This has the feeling of an electrical force, but it is too fine [an energy] to be electrical . . . Now I feel my head is separating from the body." At this, Sharon said, "That's the way I felt!" Finally, he said, "I feel the energy going into the genital area."

This ended our test. The three went off very enthusiastic, discussing how nice it would be to own such a blanket. I personally was considerably surprised at the speed and quality of their reactions. I can only assume (a) that the three were, as they appeared, very vital, free young people considerably in touch with their feelings and (b) that the weather conditions may have been a factor—it was a bright, sunny day and they sat under the blanket between two-thirty and three-thirty, which is the time of day, according to Reich, when atmospheric orgone is at its maximum power.

This account of the more interesting and varied tests made of small orgone blankets, while impressionistic, has left me with a powerful conviction that they slowly convey some energy to the body. The effects perceived typically correlate closely with the individual's sensitiveness. In the early years of usage, I felt the energy more strongly than today, some twelve years after. Nowadays I get a sense of filling up, especially if the blanket is placed over my stomach. It is easy now to tell when I've had enough.

In the early years of testing electricity, Cavendish, among

others, secured valuable measurements by using his own body "to gauge with accuracy the relative intensities of electrical shocks" (Still, p. 74). After seventeen years of continually using the orgone blanket, I believe my observations have some value, but what is needed now is rigorous scientific checks using relevant instruments. Hopefully, qualified persons will soon come forward to undertake these.

*Chapter Seventeen*

# Conclusion

In assessing the orgone theory I believe it necessary to evaluate the man and the scientific quality of his research, as well as the basics of the theory including its various implications, humanistic, scientific and philosophical.

From the standpoint of his scientific and social-philosophical contributions, I consider Reich to be an intellectual innovator and visionary of the order of Marx and Freud*—I mean someone who sees the character and problems of a society or civilization much more clearly than the average thinker and charts an extensive program to meet these problems. Moreover, owing to the brilliance of his insights, many of these ideas eventually gain some acceptance and gradually work a fundamental change in the consciousness of the civilization. Visionaries at first meet ridicule, and their ideas seldom receive wide recognition in their lifetimes. Forty years or so after publication, their core ideas slowly win allegiance from influential, though small, circles of intellectuals. After another thirty to sixty years these ideas usually have so seeped into the general educated consciousness as to turn it, albeit grad-

---

* Among well-known critics who evince a high opinion of his creativity are R. D. Laing, the British psychiatrist, Paul Roazen, the political theorist and student of Freud, and Thomas Hanna, the chairman of the Department of Philosophy of the University of Florida.

ually and partially, in a new direction. Auguste Comte, the self-confessed founder of scientific sociology, was another such visionary and in some respects Reich can be likened to him.

The speed at which a visionary's ideas become widely recognized in select intellectual circles varies, of course, with many factors, including the state of the communication media. Recent developments in media may greatly shorten the period of Reich's acceptance, although this is doubtful. Widespread intellectual recognition depends at least on the appearance of influential advocates or promoters of the ideas and on the degree to which basic socioeconomic conditions favor a hearing for the new concepts. While such well-known figures as Fritz Perls and Paul Goodman acknowledged their interest in Reich, they did not promote the orgone theory. Alexander Lowen, director of the Institute of Bioenergetics, and author of five fast-selling paperbacks* on bioenergy, is at present Reich's most widely read advocate. Yet Lowen stops short of promoting the accumulator and other core ideas and devices, or publicly seeking to persuade leading physicists to carry out scientific tests of the orgone. While the social scene in the United States is gradually turning favorable to a hearing of his ideas, the main body of physical and psychological scientists are as deaf to Reich as they were twenty years ago. In other countries, like France and Italy, some of Reich's ideas are gaining a growing audience, including radical intellectuals. It may be that he will gain the swiftest acceptance far from the shore of his adopted country, the United States.

A number of studies of creative innovators, the group from which visionaries emerge, point clearly to the emotional and intellectual qualities characteristic of such scientists. Creative scientists are typically independent thinkers, indifferent to formal religious organizations, nonconformists, strong achievers, given to insatiable curiosity, risk takers, often stubborn and hard to get along with, highly self-confident, loners except for a limited number of intellectual friendships, et cetera. Thus, an article by David McClelland, of Harvard, in H. E. Gruber, ed., *Contemporary Approaches to*

---

* These include *Love and Orgasm, Betrayal of the Body* and *Pleasure*. Drs. Baker and Raphael are the most-published orthodox Reichians.

Creative Thinking (New York: Atherton Press, 1964), summarizes various studies which conclude that creative scientists
tend to come from a background of radical Protestantism but
are not themselves religious, they tend to avoid interpersonal
contact, appear to be almost obsessed with their work, are
disturbed by complex emotions, perhaps particularly interpersonal aggression, and in their interests are intensely
masculine. They early develop a strong interest in intellectual
analysis and in the structure of things. Their goal is to get
at the inner secrets of the world (H. E. Gruber, G. Terrell,
M. Wertheimer, pp. 144–52). On such a list Reich would
score very high. Many visionaries specifically seem possessed
of a ferocious drive to excel, to discover and often to dominate;
they are intellectual imperialists. A common trait among creative scientists, says McClelland, is to dominate nature;* this
urge in many visionaries seems linked to a drive to conquer
vast territories of knowledge and subdue these to a few simple
concepts. Visionaries are obsessed by the importance of their
new concepts; some drive themselves furiously to stake out the
new territories, in their names, and while absorbing much of
the knowledge of the past, they break from it, along one or
more decisive lines. Such were Comte, Marx, Freud—so also
was Reich. Two living visionaries still churning out ideas are
Marshall McLuhan and Buckminster Fuller.

Students of human creativity emphasize that consistent
correctness in the scientific innovator or visionary, as seen
from the light of later history, is not to be expected. Correctness is, in short, not an adequate test of creative genius. In
fact, it is accepted that some errors in science, because they
raise significant questions, are in themselves creative (ibid.,
p. 44). Careful systematic testing of wrong but brilliant
hypotheses can measurably increase human knowledge. Thus,
Bacon is widely quoted for his remark, "Truth comes out of
error much more readily than it comes out of confusion"
(ibid., p. 45). In fact, just to see a new problem clearly and
to so state it, can be a really creative act. Regardless of
history's final assessment of Reich, few would deny that he
raises crucial questions about life energies—apart from his

---

* In Reich, it seems, the passion was to understand and work with nature.

brilliant contributions to psychological and sociological theory. Also, in his attempts to integrate and simplify our understanding of organisms and the cosmos, he produced fascinating theories on the harmony of nature, theories that bear the stamp of profound creativity. In terms of these kinds of considerations alone, one can justify Reich's placement among great creative scientists.

At the same time, the fact that some critics labeled Reich mad or crazy carries no weight intellectually. In point of fact, few visionaries lead what society would call a balanced life; many have been labeled queer, or unstable; quite a few were called mad; and some, like Nietzsche, did go insane. While some, like Freud and Marx, had basically happy childhoods, others, like Reich, did not. But most visionaries or radical innovators, because they are far ahead of the pack and lead a somewhat nonconformist life, are frequently criticized and labeled as dangerous, unstable or deviant by threatened vested interests and highly conventional academics. The important point is that creative visionaries place a very high premium on discovering truth, and they show high intellectual integrity, and these qualities often make them unpopular.

In general, the great strength of the visionary is that he latches on to an apparently original discovery about man, nature or society and, by dint of plowing hard and deep, digs up most of its latent explanatory powers. Seemingly totally possessed by his new conceptual tools, the visionary shows how they "light up" dark unknowns, and in the process he ignores many other factors, insights and considerations. Critics accuse him of overgeneralizing the explanatory power of his one big idea. With the passage of decades and the accumulation of new knowledge, it becomes evident the degree to which the great visionary did oversimplify, where he overgeneralized the role of his insights, how by plowing deeply his favored furrow, he blinded himself to certain facts, cultural factors and evidence. For all this, his contribution remains outstanding, and while some of his original ideas and concepts are totally rejected, and others are modified, much of his fundamental emphasis remains with its benefits to mankind. Students of the innovation process realize that, for the creative thinker, the excitement of applying the discovery widely, while usually ignoring the critics and the qualifying factors, often leads to such an unusual productivity and an impact that society is eventually forced to recognize the significance of the visionary's new concepts. Without such tenacious argument for his new ideas and for their dramatic and far-reaching import, he might well have been shoved aside or lost in the competitive struggle for innovative recognition.

It is in this total context that I view Reich's tremendous concentration on the orgone theory, its explanatory powers and his tendency toward the end perhaps to overstate its importance to mankind. He saw how much it contributed to the understanding of health and medicine, and to many natural phenomena—and a key problem was, as with most innovators, to get the world to pay attention. He probably exaggerated what the orgone could do and how effective energy devices, like the accumulator or the cloudbuster, were. I believe his later writing was overladen with repetitive and extreme claims—but in this he was not alone among visionaries. Doubtless the social and physical isolation and the persecution he met played a part. Temperamentally, he was a highly energized man, impatient, quick to anger. He was intolerant of closed minds and so he fought vigorously and near the end wrote bombastically, as befits a man deeply excited by the prospects of clarifying the incredible energy in organisms and the atmosphere. He had the energy and drive, but lacked, especially toward the end, the wisdom, calmness and subtlety to advance his ideas successfully. Much more than Freud, he needed skilled and loyal advocates—but failed to attract them in his lifetime. (Dr. Theodore Wolfe, who showed the most promise, died prematurely.) If Reich had trouble holding the loyalty of some of his followers no less did Freud or Comte before him.

Basically, Reich's personal history and his very considerable geographic and social isolation after 1950 contributed to his exaggerations and unwarranted generalizations. His childhood, while close to nature,* was lonely and particularly

---

* In the above-mentioned article Dr. McClelland hypothesizes that creative scientists "have found in nature a symbolic substitute for the idealized people they could not find in real life" (*Contemporary Approaches to Creative Thinking*, p. 163).

Conclusion												339

disturbed, since his mother committed suicide, and his father
seemed to have sought his own death. The scars from these
experiences were probably never really cleared up, since he
never completed his psychoanalysis, after one of his analysts
abruptly left the country. This may have contributed to an
excessive amount of hostility to authority symbols; one
wonders if he did not unconsciously seek out, or even invite,
trouble and persecution from traditional institutions. Then, by
1945, after his many put-downs, he abandoned the regular
scientific community and thus eschewed both its potential social support and its constructive criticism. In isolated Rangeley,
Maine, he worked hard to understand and develop the potentials of the orgone theory and gave up all political action.
Later, by concentrating *without the necessary resources* on the
huge problems of controlling nuclear radiation, the growth of
deserts and the emotional plague, he demonstrated a serious
lack of balance. But imbalances are common among visionaries; and according to the reports of friends Reich had many
moments of lucidity and wisdom right up to his death. Unques-
tionably, Reich earned the label of genius, but with all his insights he was subject to errors of intellectual judgment, of
intuition and of scientific procedure—like other visionaries.
Utopian in some of his thinking, he excited unrealistic utopian
hopes in some followers. In addition, the easily sensationalized
nature of some of his ideas on sex and his claims for the orgone left him an easy prey to irresponsible journalistic criticism or attacks. And from this it was but a short step to more
serious persecution.

From today's perspective it is easy to condemn, as Ralph
Nader does, the concerted attacks made on him by the Food
and Drug Administration (James S. Turner, *The Chemical
Feast*, pp. 32–33), but how many scientists profoundly concerned with conquering the cancer scourge would have taken
such persecution calmly? Reich got angry, fought back rather
irrationally, and in the process apparently developed paranoid
symptoms. Perhaps he was very naïve not to expect some
persecution for his radical notions; but one has to expect
some shortcomings in great men who, like Reich, become
obsessed with scientific research. Their very concentration
on research findings may leave them simple-minded or vulner-

able in other areas. And many are unable to handle aggression. Dr. McClelland has established that creative scientists have more than normal problems handling frustration and aggression (*Contemporary Approaches to Creative Thinking*, p. 172). This was clearly true of Reich, whose violent anger antagonized more than one brilliant co-worker and whose writings, especially after 1945, often seem designed to disturb, shock or hurt.

Throughout his life Reich repeatedly labeled himself a natural scientist and placed himself in the tradition and line of the great natural scientists from Galileo and Bruno through Newton and Semmelweis to Freud. His goal in life was to probe the secrets of nature for the benefit of mankind. Like the great masters of the past he worked alone or with a few assistants, in a small personally equipped laboratory, often short of funds to purchase necessary equipment. He wrote:

> Some readers may ask why this or that experiment was not also carried out, why this or that substance was not also investigated. I am painfully aware of many such gaps. They are, however, not my fault, but largely that of the circumstances under which the work had to be carried on. Several academic organizations which could have lent their financial support considered the whole thing, on the basis of individual, unrelated findings, so fantastic that I had to decide to maintain my independence and to get along without outside help. . . . The experimental work consumed practically my whole income; during the years of 1940 and 1941 alone, it cost more than $10,000. Adequate research facilities would have required hundreds of thousands of dollars. The limitation in the use of materials, apparatus and experimental setup simply reflects the limits of my economic capacity. [*Selected Writings*, p. 188]

Thus Reich eschewed the whole concept of team and large-grant-supported research. While this is atypical in the late-twentieth century, it is historically the way in which all the great scientific innovators of the past carried out their work.

In such rather lone-wolf research, everything depends on the methodological skill, the basic scientific knowledge and integrity of the researcher, and his capacity to learn from mistakes. On most of these counts Reich stands up well to criticism. He was well grounded in the scientific tradition, through his medical training, a ten-year association with the Freudian circle in Vienna, and his wide reading. He clearly was prepared to and did learn both from mistakes and from casual but penetrating observations of nature; his scientific integrity was unimpeachable. Where he might be faulted was in his methodology of research and particularly his writing up of results in articles and books on the orgone.

It may be useful to outline a number of fairly obvious criticisms. Firstly, one should note the strong influence of Reich's training in psychoanalysis, where conclusions are drawn from intensive study of a few cases of pathology, plus day-to-day clinical observations. This procedure may be acceptable in launching speculative theories in a wholly new field like psychoanalysis, but in tested sciences like biology and physics it is a risky procedure, calculated to arouse suspicion or ridicule.

Secondly, owing to the paucity of funds and the difficulty of constructing controlled, repeatable experiments, Reich carried out little replication or testing of his original hypotheses. It is true, he tested the orgone on cancerous mice in controlled experiments. However, for lack of funds, equipment and assistants these and other controlled tests were rarely repeated more than a few times and hence do not satisfy present-day scientific expectations. Again, in his work with humans—for example, in using the accumulator—Reich tended to rely upon a few cases, seldom repeated or fully controlled, with the result that the influence of extraneous variables could not be said to be unequivocally eliminated. Thirdly, the weather-control experiments were at best pilot experiments, exploring a possible line of inquiry, rather than yielding "hard" facts. Yet, in his writing, especially in the latter years, Reich tended to claim important findings out of such pilot endeavors and occasionally generalized rather grandiosely. At other times he admitted "there are many gaps to be filled."

Fourthly, throughout much of his scientific writing after 1945, and occasionally before this, he failed to write up his research in the precise, cool and detached manner expected of scientists. Instead, Reich provided very brief explanations*

---

* He admitted in one place, "I could have presented the *Function of the Orgasm* in 1,000 instead of 300 pages and the orgone therapy of *The*

and graphics, failed to qualify the results, often rambled on to ideas or theories associated only mentally with the original experiment, and sometimes presented his material in the excited, declamatory and confident manner reminiscent of a young college student highly elated over early scientific "discoveries." Orthodox scientists were accordingly horrified and very critical.

Reich's methodology must be seen as closely related to certain factors in his training, social experience and psychic constitution. He had great confidence in his ability to grasp essential relationships in nature, a confidence bred by his early boyhood biological investigations (on the farm) and his skill in gaining the acceptance of the Freudian circle in Vienna. At the age of twenty-seven—ten to fifteen years younger than almost all the rest—he was made Director of the Seminar for Psychoanalytic Therapy, and at thirty-one (in 1928) vice-director of the Polyclinic; in addition, his early papers and books were well received in psychoanalytic circles. Thus, he was early "trained" to fight for recognition and for his ideas and to see himself as precocious. My hunch, furthermore, is that an impressive source of Reich's discoveries of natural relationships was an unconscious but unusual sensitivity to emotional states bordering at times on clairvoyance. Reichians might call it "orgonotic" contact. He just "knew" that certain things were so, and hence failed to see the necessity of careful, precise reporting of supporting data and experiments. A number of other innovators seemed to have gained many of their fundamental discoveries by somewhat similar intuitive processes.

An interesting article by psychologist David Bakan, "Science, Mysticism and Psychoanalysis," may assist our evaluation of Reich's work. Bakan maintains that all scientists, in their perspective, can be plotted on a continuum of mysticism and that some of the greatest of them revealed strong mystical leanings—for example, Newton, Kepler, Pascal, Einstein, Schrödinger and Max Planck. High in mystical leanings are

---

*Cancer Biopathy* in 5,000 instead of 500 pages. I admit further that I never made the effort to familiarize the reader with the essence of my method of thinking and doing research . . . this has undoubtedly done much harm" (*Selected Writings*, p. 21).

those mathematicians who are attracted to the discipline by its novelty, promise of excitement and "admission to wild and unaccountable spaces" (Bakan, "Science, Mysticism and Psychoanalysis," p. 1). Quoting Bertrand Russell's *Mysticism and Logic*, Bakan notes that "the refusal to admit opposition or division anywhere," and a conviction that that which is not manifest influences that which is manifest, characterize the mystical style. The conviction of a basic unity behind apparent disunity has been the key theme of much science; indeed, Bakan sees science itself as potentially mystical. Moreover, the drive of psychoanalysis to make manifest the forces of the unknown unconscious underscores an implicit appeal to the mystical-minded. (Bakan has explored some of this in his *Sigmund Freud and the Mystical Jewish Tradition*.) According to this line of reasoning, the orgone theory, especially in its cosmic dimensions, places Reich on the mystical left.

Psychoanalyst Charles Rycroft says Reich was but a nature poet struggling to escape his scientific armoring. He sees the orgone theory as a pseudo-scientific version of Blake's assertion that "the innocent [i.e., the unarmored] can 'see a world in a grain of sand and a heaven in a wild flower' " (Rycroft, p. 85). From Bakan's viewpoint, such a judgment is typical of the nonmystical scientist judging the mystical; it does not invalidate the legitimacy of Reich's efforts or the orgone hypothesis. In fact, says Bakan, since history shows that innovators are more likely to be on the mystical left—that is, high in mysticism—and to sway those on the right of the continuum, Rycroft's remarks can be seen as subjective reflections of his antimystical stance.

In assessing Reich's methodology it is necessary to recall his long apprenticeship to psychoanalysis and his strong and admiring relationship to Freud. This provided him with a training in clinical discovery, in which single exceptional cases or a few unusual cases occasionally launched, or *seemed* to indicate, a new range of theory. Freud and his close followers worked this way. As an example, consider the extensive theorizing that Freud and others did on the Wolf-Man case. For a recent discussion of this, there is Muriel Gardiner's *The Wolf-Man, by the Wolf-Man*. Paul Roazen provides a very thoughtful review of the book in *Saturday Review*,

August 14, 1971. In general, the Freudians were probing a whole new understanding of human action, and they were tremendously excited by the theoretical possibilities of unusual case findings. They seem to exhibit something of the rather common German compulsion to theorize and generalize. Moreover, to defend the importance of their innovative work —which gave deep meaning to their lives—and to counteract the criticism and ridicule of the medical establishment during the 1910's and 1920's, they theorized extensively and rabidly. Reich was part of this during his intellectually formative years from twenty-four to thirty-six, and so "inherited" a specific intellectual life style and rather naturally carried on in this tradition.

Furthermore, the Freudian tendency to make a lot out of a few unusual cases, to generalize radically about human nature from clinical findings, must also be seen as related to the religious positions of both the Freudians and of Reich. All were, with a few exceptions, Jews who had abandoned Judaism and needed a new religion to give focus and meaning to their lives. And their social marginality as Jews and intellectual innovators increased the tendency of making psychoanalysis (or, in Reich's case, orgonomy) into a religion, a system of ultimate understandings of human and inanimate nature. So Reich's tendency to deify life, with a capital L,* and his extensive and often metaphysical generalizing probably flowed from this need to find and affirm a socioreligious perspective and so to fill the vacuum of an absent theism. The dogmatism of the various Freudian "sects" and latterly of the growing variety of Reichian sects is undoubtedly linked to the same situation: ex-Christians and ex-Jews needing to find a new and true religion. (No one is more dogmatic than the new convert, especially the convert to a new or unpopular faith.)

Reich, in his later years, elaborately rationalized his departure from the scientific canons of the late-twentieth century. He roundly castigated many scientific procedures and findings as mechanistic and inappropriate, and said they reflected armored feelings. He claimed that one should freely

---

* Philip Rieff in his *The Triumph of the Therapeutic* has ridiculed this trait in Reich without trying to understand its source.

express one's feelings, in science and life, not hide or repress them, and he criticized normal science as largely a mechanistic enterprise of little value to humanity. In part, Reich maintained a Marxian position that science cannot be value-free or really detached, and that those who make such claims are only fooling themselves and/or the public. Today a good deal of social and some biological science is gradually polarizing between the quantitative and sociopolitically "neutral" and the qualitative, sociopolitically radical; it is likely that some in the latter group will be attracted to and find much to admire in Reich's work and his style of science.

Obviously the value of the more intuitive and qualitative methodology typified by Reich or Freud depends heavily upon the sheer brilliance of the scientist himself. Almost anyone with some training can engage in quantitative team science and get some results that will stand up to rigorous scrutiny, whereas the long-term value of the qualitative or clinical investigator depends heavily upon his innate capacity. Moreover, persons like Comte, Marx, Freud or Reich (or McLuhan) may be penetrating and brilliant in some areas, or at the early periods of their life, but show weakness and erratic judgment in other cases and later periods. Also, the more intuitive methodology is much faster at producing "conclusions" than the ponderous quantitative approach and raises greater hopes and expectations. By the same token it may be far wider of the mark.

In general, it is my considered opinion that Reich's methodology was strongest in throwing light on broad human and natural relationships. It opened up suggestive connections regarding the possible linkages of life energies to body armoring, the emotional basis of cancer, and to weather and drought conditions. But this very strength may also have a weakness—that is, that specific "findings" may well turn out, after concentrated replicated experimentation and general scientific advance, to be misinterpretations.

At times Reich was willing to admit he was charting out whole new territories only in broad compass. At one point he compared his work to that of Columbus who discovered a barren coastal stretch on the Atlantic Ocean, but none of the treasures of the new land. He says: "I have in reality made only one single discovery: the function of the orgastic-plasma contraction. It represents the coastal stretch from which everything else has developed" (*Selected Writings*, p. 19). Yet, at other times, usually in his later days, he made very strong imperialistic claims. Some neo-Reichians like Charles Kelley have been more circumspect. Yet in the nature of the case the most that one would expect from any human working with the inadequate facilities that Reich possessed is that, like Columbus, he would roughly chart out new and exciting approaches to knowledge and that other men will spend decades clarifying them, correcting here and there, and bringing them into more precise fit with gathered facts.

Some critics may want to evaluate the orgone theory directly in terms of its value for therapy. Reich was less and less concerned after 1940 with devising a totally adequate therapeutic armamentarium. His interest in cancer and schizophrenia and the degenerative diseases was more theoretical than therapeutic. His methodology was focused on turning up fundamental relationships in nature and not with the minutiae of therapeutic exactitude. His aim was to show men how to attain, in broad terms, their lost animal vigor, health and dynamism—and even here his methodology can be described as largely tangential and exploratory. If he ignored key concerns and issues of ego and existential psychiatry as he did, it was because they were of no interest to him. Such issues belonged to the field of psychology and psychoanalysis, which he had deliberately abandoned by at least 1940. Some will see this almost total focus on energy considerations, to the neglect of a more humanistic profound psychology, as a serious weakness. Some critics see the concentration on the orgone as a mechanistic or superficial approach to therapy. But this is to misread the man's genius—to keep moving into nature's secrets. The excursion into vegetotherapy was brief, though insightful. Ellsworth Baker has extended this, for example, in his treatment of the eye segment. Alexander Lowen and the bioenergetic group have picked up and used certain key ideas while developing some of their own therapeutic insights. In the same way, but more slowly, I believe we will see other of Reich's therapeutic interests—for example, with young infants—expanded and specified.

I am convinced in broad outline that much of what Reich

said about the orgone is correct. In terms of its precise physical properties, I suspect that the energies finally pinpointed will be more complex than Reich himself claimed. What has to be reiterated is that no reputable scientist or team of scientists has taken the time to duplicate the basic orgone experiments to find whether they are sound or erroneous. The time must soon come when some young scientists will repeat the basic experiments and announce their findings.

I believe the evidence for a radiational force with its main source in sun or star energy, linked with the atmosphere and conducted through living organisms, is now impressive. I view this energy as basic to life; however, I have doubts that it is primordial to the whole universe, as Reich latterly claimed. This judgment seems premature to me. I am certain this energy (the orgone) has certain healing and vitalizing properties when properly conveyed to organisms. In its nature it may be purely electrical, as we now understand electricity, or closely akin to electricity. I leave that judgment to the physicists. But I believe that it may be related to negative ions; although precisely how, only future research will reveal. It seems clear that this energy produces a field around human and other organisms, that this field is sensitive to microwaves of various sorts and that it varies with the emotional state of the person or animal. Under the right conditions certain persons are able to project a part of the field force some distance from the body, and thereby carry out a healing function. The use of the hands may assist such energy projection, but their use is not essential to its healing utilization.

There are ways of concentrating this organismic energy for human or animal use, and I believe that the accumulator is but one of these ways. I suspect that Reich and his close followers may have overstated the physical benefits of using the accumulator; on the other hand, I am confident that its mild healing powers could be of real importance to mankind and that eventually they will be exploited. Therapeutically, I believe the critical question is not the existence of the orgone, but how best the human organism can utilize it. Merely "pumping it in" via the accumulator is not an answer to many complaints; various combinations of it, along with bioenergetic therapy and maybe other processes, are needed.

What of the precise characteristics of the orgone and of

its more specific capacities? Here, my private view is that Reich was probably as often wrong as he was right. I am not convinced that the orgone is not polarized, nor that Reich's theory as to how it is "collected" by the accumulator is totally correct. Again, his concept of orgone stasis in the body and its connection with body armoring impresses me as a rough beginning insight, only partially accurate. Reich's ignorance of studies about the body's electrical force field (Burr and Ravitz) and the work of Tromp, Maby and others on radiesthesia and biometeorology, led him incorrectly, I think, to see organismic energy as a single, unitary force, quite distinctive from electromagnetic phenomena. I believe that further research will eventually demonstrate a variety of energies now unknown to orthodox science and that some of them will be identified or related to what Reich called orgone. Work in psychotronics being carried out in Russia, Czechoslovakia and Britain, and reported in the *Journal of Paraphysics*, is but one type of research that tends to support this belief.

I am similarly equivocal about the absolute accuracy or validity of related aspects of the orgone theory. Thus, while I accept the role of orgone blockages in the generation of many, but by no means all, cancers and other degenerative diseases, I am not convinced that Reich pinpointed the exact relationship between emotion and disease. Its complexities have yet to be more fully unraveled. The therapeutic value of vegetotherapy is considerable, and I accept that its latest version, bio-energetics, can produce, in skilled hands, some therapeutic marvels. There are limitations to what it can achieve by itself apart from the fact that its goal is restricted to animal energy efficiencies and pleasure sensations. To what extent severely disturbed individuals, or persons beyond the age of thirty-five or forty, can profit greatly by psychiatric orgone therapy or by bioenergetics has yet to be clarified. Also, how many persons can be brought to orgastic potency and to what extent that goal is (a) worth the effort, for older or disturbed persons, or (b) is simply stage one in a longer and more spiritual pilgrimage, I am not certain. But I am impressed by those who point out that Reichian therapists do not seem *as a group* to have achieved, any more than Reich did, an exalted or enviable level of personal equanimity or inner happiness.

The Oranur experiment and Reich's theories on DOR, MELANOR (the material form of DOR) and weather control strike me as ingenious attempts to come to grips with very complex phenomena rather than as simply dubious endeavors. Clearly Reich was qualified to innovate only in the broadest terms here, and one must lean heavily on persons like Charles Kelley, who did have meteorological training. I accept the existence of toxic qualities of organismic energy, for I've seen some of its effects; on the mechanism of its operation and Reich's ways to "control" it, I adopt a wait-and-see stance. Much research must be done before a clear picture will emerge.

Reich saw many of the problems inherent in his kind of weather changing. If he is proved even partly correct, the field of weather control in even one nation will pose grave problems for the international community. Over-all and particularly in connection with DOR and the weather-control work, my opinion is that Reich stumbled upon a tremendous natural force whose powers staggered his imagination, and that while he tried to define its parameters and capacities, he didn't take time for—or show much interest in—discovering its limitations, physical, therapeutic, meteorological or social.

When it comes to extensions and implications of the orgone theory, as seen in such theories as cosmic superimposition, orgonotic functionalism or Reich's views on political and social developments, I am both intrigued and impressed. I see these extensive theories as great stimulants both to scientific theory and speculation and to man's attempt in these critical days to appreciate better our complicated, ever-growing universe and our place within it.

The philosophic, sociological and cosmic implications of the orgone theory are for me and many others extremely stimulating. For some time Reich has been quoted here and there for his libertarian views on sexual freedom, on the necessary demise of the authoritarian family, on women's liberation, on the need for a pleasure-centered, instead of pleasure-denying culture, on the tremendous importance of surrounding the infant, before and after birth, with love and

freedom, on the peripheral importance of politics in any lasting social change, and on the tremendous significance of sexual
happiness for emotional and social health. As the ecological
and whole-earth movements gain strength, he will be quoted,
perhaps frequently, as the visionary who dared to place man
in intimate emotional relationship with the whole of nature
including the atmosphere, the weather and cosmic forces.
In all these areas Reich functioned as a Jewish prophet who
like Marx hammered away at the antilife and antihuman
aspects of an overripe civilization. Well in advance of more
popular thinkers like Ellul, he castigated the mechanistic
emphasis of modern society and its knowledge systems.
Basically, Reich represents a return to a rather primitive
metaphysics, emphasizing the centrality of the sexual encounter, man's orgonotic roots in nature, the role of energy
in personal and social transactions. He calls alienated civilized
man to remember his lost roots, including the primitive intuition of one's mystical affinity with all of nature. As man's
estrangement from an overly mechanistic civilization and
from his own self accelerates, more and more people will
likely read and reflect on Reich's ideas. And as these seekers
for new life turn away from a sycophant scientific establishment tied to scandalous military destructiveness, unorthodox
scientists like Reich will gain in popularity.

Dr. Murray Melnick of Hofstra University recently has
written to me:

> I believe that Reich represents a thrust into the future—and
> that the poetic intensity of his work is just too much for some
> people to bear, psychologically. Our science, our civilization
> leads a nice, sure, controlled understood life—the truly un-
> known and inexplicable is feared; and yet all life, all real love
> has its awesome quality! To *feel* what Reich means is to won-
> der whether there is not something beyond all mechanical, life-
> less and unloving measurements.
>
> There is conveyed in his work the feeling that Life is bigger
> and grander than it seems—there is a longing for deep related-
> ness to all existence. Reich is the hope that things are better
> than the petty meanness of our life.
>
> The fears I mention may be at the core of much of the re-

sistance to Reich; the hopes genuine enough but this does not prove that Reich is right. The validity of the orgone must ultimately be decided by scientific means. It cannot be disposed of either in the affirmative by pointing out the reasons for resistance or in the negative by persistent refusals to test amidst concurrent unsupported assertions *ad nauseam* that Reich has been disproved, Reich is wrong, Reich is crazy.

I see numerous specific ramifications to Reichian writings and the orgone theory. As Thomas Hanna points out in *Bodies in Revolt*, Reichian emphases will assist in restoring the body to its proper place in Western thought. The orgone theory, tying in with yoga and acupuncture perspectives as it does, will assist in the coming blend of Oriental and Western religiosity and philosophy. We can also expect that the orgone theory will play a part in the emergence of a new sexual morality—stressing honest giving and receiving of affection—the acceptance of uninhibited sexual response and the saliency of creative work of all kinds. As the Western world moves away from the patriarchal authoritarian family and traditional ascetic approaches to sex and the body, sexual license will increase along with some genuine advance in sexual freedom. Reich foresaw and deplored the trend to pornography; his theory helps to account for sexual excesses and in many ways, by his honesty and his freeing of the emotions, has intellectually paved the way for a revolutionary approach to the sexual encounter and to new family structures. Eventually our society will work out a new ethic of play, work and family life, much of which may well bear heavily the imprint of Reich's thinking.

Specifically, I expect the orgone theory to provide some basic guidelines for radical new directions in medicine, psychosomatics and psychiatry. Hospitals as institutions will be late to react, but in time they must respond. Much of the coming revolution in psychotherapy will likely result from the encounter-therapy movement and its increasing tendency to favor bio-energetics and freer perspectives on the body. Further, I fully expect that more experimenters in the ESP field, following up the advances made (re the bioplasms) in Soviet-bloc nations, will "discover" the orgone theory and bio-energetic therapy and find them extremely useful both theoretically and in opening up many persons to latent psychic capacities.

Reich's ideas excite the scientific imagination. They presen us with an organic and holistic approach to nature desperatel needed at this critical juncture. His orgone theory has createc valuable intellectual controversy and no doubt will continue to do so. Certain elements in Western society will be fairly quick to adopt some of his ideas, a few groups will experiment with the orgone, and cultlike movements, gaining a foothold in certain socially marginal areas, may treat Reich as another Christ. The therapeutic implications of the theory will doubtless produce valuable spin-offs. I doubt that Reich or the full orgone theory will gain wide acceptance soon. But it may well play a very significant part in the wave of the future.

# APPENDIX A

## SCIENTIFIC DEVELOPMENT OF WILHELM REICH

Wilhelm Reich's basic scientific discoveries include the following: orgasm theory and technique of character analysis (1923–34); respiratory block and muscular armor (1928–34); sex-economic self-regulation of *primary* natural drives in their distinction from *secondary, perverted* drives (1928–34); The role of irrationalism and human sex economy in the origin of dictatorship of all political denominations (1930–34); the orgasm reflex (1934); the bioelectrical nature of sexuality and anxiety (1935–36); orgone-energy vesicles, bions (1936–39); origin of the cancer cell from bionously disintegrated animal tissue, and the organization of protozoa from bionously disintegrated moss and grass (1936–39); T-bacilli in sarcoma (1937); discovery of the bioenergy (orgone energy) in SAPA bions (1939), in the atmosphere (1940); invention of the orgone-energy accumulator (1940); and the orgone-energy field meter (1944); experimental orgone therapy of the cancer biopathy (1940–45); experimental investigation of primary biogenesis (Experiment XX, 1945); method of orgonomic functionalism (1945); emotional plague of man as a disease of the bioenergetic equilibrium (1947); orgonometric equations (1949–50); hypothesis of cosmic superimposition of two orgone energy streams as the basis of hurricanes and galaxy formation (1951); antinuclear radiation effects of orgone energy (The Oranur Experiment, First Report, 1947–51); discovery of DOR (deadly orgone energy) and identification of its properties, including a specific toxicity (DOR sickness) (1951–52); identification of Melanor, Orite, Brownite and Orene, and initial steps toward preatomic chemistry (1951–54); use of

353

Appendix A

"reversed" orgonomic potential in removing DOR from the atmosphere in cloudbusting and weather control (1952–55); theory of desert formation in nature and in man (the emotional desert) and demonstration of reversibility (Orop Desert Ea and the Medical DOR-Buster) (1954–55); theory of disease based on DOR accumulation in the tissues (1954–55); equations of gravity and antigravity (1950–57); development and practical application of social psychiatry (1951–57).

# Appendix B

ANORGONIA. The condition of diminished or lacking orgonity (q.v.).

ARMOR. *See* character armor, muscular armor.

BIONS. Energy vesicles representing transitional stages between non-living and living substance. They constantly form in nature by a process of disintegration of inorganic and organic matter, which process it has been possible to reproduce experimentally. They are charged with orgone energy (q.v.)—i.e., *Life Energy*— and may develop into protozoa and bacteria.

CHARACTER. An individual's typical structure, his stereotype manner of acting and reacting. The orgonomic concept of character is functional and biological, and not a static, psychological or moralistic concept.

CHARACTER ARMOR. The sum total of typical character attitudes, which an individual develops as a blocking against his emotional excitations, resulting in rigidity of the body, lack of emotional contact, "deadness." Functionally identical with the muscular armor.

EMOTIONAL PLAGUE. The neurotic character in destructive action on the social scene.

MUSCULAR ARMOR. The sum total of the muscular attitudes (chronic muscular spasms) which an individual develops as a block against the breakthrough of emotions and organ sensations—in particular, anxiety, rage and sexual excitation.

ORGASM. The unitary involuntary *convulsion of the total organism* at the acme of the genital embrace. This reflex, because of its *involuntary* character and the prevailing orgasm anxiety, is blocked in most humans of civilizations which suppress infantile and adolescent genitality.

ORGASTIC POTENCY. Essentially, the *capacity for complete surrender to the involuntary convulsion* of the organism and *complete discharge* of the excitation at the acme of the genital embrace. It is always lacking in neurotic individuals. It presupposes the presence or establishment of the genital character—i.e., absence of a pathological character armor and muscular armor. Orgastic potency is usually not distinguished from erective and ejaculative potency, both of which are only prerequisites of orgastic potency.

ORGONE ENERGY. Primordial Cosmic Energy; universally present and demonstrable visually, thermically, electroscopically and by means of Geiger-Müller counters. In the living organism: *Bioenergy, Life Energy.* Discovered by Wilhelm Reich between 1936 and 1940.

ORANUR denotes orgone energy in a state of excitation induced by nuclear energy. (DOR denotes *Deadly OR* energy.)

ORGONOMETRY. Quantitative orgonomic research.

ORGONOMIC ("ENERGETIC") FUNCTIONALISM. The functional thought technique which guides clinical and experimental orgone research. The guiding principle is that of the *identity of variations in their* common functioning principle (CFP). This thought technique grew in the course of the study of human character formation and led to the discovery of the *functional* organismic and cosmic orgone energy, thereby proving itself to be the correct mirroring of both living and non-living basic natural processes.

ORGONOMY. The natural science of the cosmic orgone energy.

PHYSICAL ORGONE THERAPY. Application of physical orgone energy concentrated in an orgone energy accumulator to increase the natural bioenergetic resistance of the organism against disease.

SEX-ECONOMY. The body of knowledge within Orgonomy which deals with the economy of the biological (orgone) energy in the organism, with its *energy household*.

# APPENDIX C

DEPARTMENT OF
## HEALTH, EDUCATION, AND WELFARE
FOOD AND DRUG ADMINISTRATION
WASHINGTON 25, D. C.

April 30, 1957

Mr. W. E. Mann
10 Veery Place
Don Mills, Ontario
Canada

Dear Mr. Mann:

Your letter of April 23 requested the names of the expert clinicians and research personnel who tested the Orgone Energy Accumulator devices for the Government in connection with the Wilhelm Reich case.

Since none of these experts testified in the court proceedings, it is regretted that we are unable to provide you with any information concerning them. It is only when such experts testify in court and their testimony becomes a part of the court record that we release information as to their names, addresses, and testimony.

Sincerely yours,

K. L. Milstead, Director
Division of Regulatory Management
Bureau of Enforcement

# BIBLIOGRAPHY

Abetti, Giorgio, *Solar Research*. New York: Macmillan, 1963.
Adamenko, Victor, *Electrodynamics of Living Systems*. Unpublished translation from the Russian original issued by Novosti Press Agency.
Alrutz, S., "Une Nouvelle Espèce de rayonnement de l'organisme humain," *Archives de neurologie et psychiatrie*, Zurich, 1922.
Angoff, Allan, *The Psychic Force*. New York: Putnam, 1970.
Ash, Michael, *Health and Radiation*. Enfield, Middlesex: ARO Enfield, 1955.
————, *Health, Radiation and Healing*. London: Darton, Longman and Todd, 1962.
Bach, Christian, *Ions for Breathing*. New York: Pergamon Press, 1967.
Bagnall, Oscar, *The Origin and Properties of the Human Aura*. Secaucus, N.J.: University Books, 1970.
Bahnson, C. B., "In Memory of Dr. David Kissen: His Work and His Thinking," *Annals of the New York Academy of Sciences*, Vol. 164, Art. 2, 1969.
————, and Bahnson, Marjorie B., "Ego Defenses in Cancer Patients," *Annals of the New York Academy of Sciences*, Vol. 164, Art. 2, 1969.
Bakan, David, *The Duality of Human Existence*. Boston: Beacon Press, 1966.
————, "Science, Mysticism and Psychoanalysis," *Catholic Psychological Record*, Vol. 4, No. 1 (Spring, 1966).
Baker, Ellsworth, *Man in the Trap*. New York: Macmillan, 1967.
Barety, A., *Le Magnétisme animal étudié sous le nom de force nervique rayonnante et circulante*. Paris: Doin, 1887.

Barnothy, Madeleine F., ed., *Biological Effects of Magnetic Fields.* New York: Plenum Press, 1964.

Barth, Lawrence, "Orgone in Relation to Some Other Energies," *Creative Process,* June, 1965.

Bazett, L. Margery, *Beyond the Five Senses.* Oxford: Basil Blackwell, 1946.

Bean, Orson, *Me and the Orgone.* New York: St. Martin's Press, 1971.

Benet, Alfred, and Fire, Charles, *Animal Magnetism.* New York: Appleton, 1888.

Bennette, Graham, "Psychic and Cellular Aspects of Isolation and Identity Impairment in Cancer: A Dialect of Alienation," *Annals of the New York Academy of Sciences,* Vol. 164, Art. 2, 1969.

Binstoc, O., "Therapeutic Value of Microwaves," *International Conference of Physical Medicine.* Amsterdam: Excerpta Medical Foundation, 1964.

Bisher, R. O., Bachman, C. H., and Friedman, H., "Is Magnetism the Key to Life's Origin on Earth?" *Saturday Review,* Vol. 46, No. 27 (July 6, 1963).

————, "The Research Frontier," *Saturday Review,* Vol. 45, No. 5 (Feb. 3, 1962).

Bizzi, Bruno, "Orgone Energy: Life Force and Morbid States," *Energy and Character,* Vol. 1, No. 1 (Spring, 1970).

Black, Stephen, *Mind and Body.* London: Kimber, 1969.

Blasband, Richard A., "Orgonomic Functionalism in Problems in Atmospheric Circulation," *Journal of Orgonomy,* Nov., 1969; May, 1970; Nov., 1970.

Boadella, David, ed., *Energy and Character,* Spring, 1970; September, 1972.

Boddington, Henry, *The Human Aura and How to See It.* London: Psychic Press, 1933.

Boirac, Émile, *Our Hidden Forces.* New York: Stokes, 1917.

Boischot, A., *Le Soleil et la terre.* Paris: Presses Universitaire de France, 1961.

Booth, Gotthard, "General and Organic-Specific Object Relationships in Cancer," *Annals of the New York Academy of Sciences,* Vol. 164, Art. 2, 1969.

Bromberg, Walter, *The Mind of Man—History of Psychotherapy and Psychoanalysis.* New York: Harper, 1963.

Brown, F. A., *Biological Clocks.* Boston: American Institute of Biological Sciences, 1962.

————, "Extrinsic Timing of Rhythms," *Annals of the New York Academy of Sciences, 1962,* p. 775.

Brown, Malcolm, "An Introduction to Direct Body-Contact Psychotherapy." Privately published, 1971.

Brunler, Oscar, *Rays and Radiation.* Los Angeles: De Vorss, 1950.

Bunning, E., *The Physiological Clock.* Berlin: Springer, 1964.

Burr, Harold S., "Electrometrics of Atypical Growth," *Yale Journal of Biology and Medicine,* Vol. 25 (1952–53).

————, *Blueprint for Immortality*. London: Neville Spearman, 1972.
Carrington, Hereward, *Higher Physical Development (Yoga Philosophy)*. New York: Dodd, Mead, 1920.
Cayce, Edgar, *Auras*. Virginia Beach, Virginia: ARE Press, 1945.
Cloudsley-Thompson, J. L., *Rhythmic Activity in Animal Physiology and Behaviour*. New York: Academic Press, 1961.
Coblenz, W. W., *Man's Place in a Superphysical World*. New York: Sabian Publishers, 1954.
Cody, P., *Sur la Radiation du sol*. Le Havre, 1933.
Colquhoun, John Campbell, *Report on Experiment on Animal Magnetism to the Edinburgh Society*. Edinburgh: Robert Cadell, 1833.
Constable, Trevor, "Operation Backwash, #1," *Journal of Orgonomy*, Nov., 1971.
Coronili, S. C., *Problems of Atmospheric Space Electricity*. Amsterdam: Elsevier, 1965.
Couteneau, G., *La Divination chez les Assyriens et les Babyloniens*. Paris: Payot, 1940.
Croon, Richard, *Elektroneural-Diagnostik und Therapie*. Baden: Verlag Konkordia, 1949.
Darnton, Robert, *Mesmerism and the End of the Enlightenment in France*. Cambridge, Massachusetts: Harvard University Press, 1968.
Debus, Allen G., *The English Paracelsians*. New York: Franklin Watts, 1965.
Doherty, Beka, *Cancer*. New York, 1949.
Eden, Jerome, *Our Planet Is in Trouble*. Valdez, Alaska: Eden Press, 1959.
Eeman, L. E., *Cooperative Healing*. London: Frederick Muller, 1947.
*Electronic Medical Digest*, Autumn, 1950.
————, Second Quarter, 1953.
Ellinger, F., and Lanz, M., *Biologic Fundamentals of Radiation Therapy*. New York: American Elsevier, 1941.
Faraday, Ann, "The Return of Reich," *New Society*, Sept. 3, 1970, pp. 404–6.
*Fifth Spiritual Healing Seminar*. Rye, New York: Laymen's Movement for a Christian World, 1956.
Frost, Evelyn, *Christian Healing*. London: M. Parrish, 1955.
Gallert, Mark, *New Light on Therapeutic Energies*. London: James Clarke, 1966.
Gamow, George, *A Star Called the Sun*. New York: Viking, 1964.
Gardiner, Muriel, *The Wolf Man, by the Wolf Man*. New York: Basic Books, 1971.
Garrett, Eileen J., *Awareness*. New York: Berkeley Publishing, 1968.
Garrison, Omar V., *Tantra: The Yoga of Sex*. New York: Julian Press, 1964.
Gasc-Desfossés, Édouard, *Le Magnétisme vital*. Paris, 1907.

Gauquelin, Michel, *The Cosmic Clocks.* Chicago: Henry Regnery, 1967.
————, *Scientific Basis of Astrology.* New York: Stein and Day, 1969.
Glasser, Otto, *Medical Physics.* Chicago: Year Book Medical Publishers, 1948.
Goldsmith, Margaret L., *Franz Anton Mesmer.* Garden City, New York: Doubleday, 1934.
Grad, Bernard, "Healing by the Laying on of Hands: Review of Experiments and Implications," *Pastoral Psychology,* September, 1970, pp. 19–26.
————, "A Telekinetic Effect on Plant Growth," *International Journal of Parapsychology,* Vol. VI, No. 4 (1962).
————, Cadoret, R., and Paul, G. I., "An Unorthodox Method of Treatment of Wound Healing in Mice," *International Journal of Parapsychology,* Vol. III, No. 2 (1961).
Gruber, Howard E., *Contemporary Approaches to Creative Thinking, A Symposium.* New York: Atherton, 1964.
Guallierotti, R., "The Influence of Ionization on Endocrine Glands," *Bioclimatology, Biometeorology and Aeroionotherapy.* Milan: Carlo Erba Foundation, 1969.
Halacy, Daniel S., Jr., *Bionics, the Science of "Living" Machines.* New York: Holiday House, 1965.'
————, *Radiation, Magnetism and Living Things.* New York: Holiday House, 1966.
————, *Weather Changers.* New York: Harper & Row, 1968.
Hansen, K. M., "Some Observations with a View to the Possible Influence of Magnetism on the Human Organism," *Acta Medica Scandinavia,* Vol. 97 (1938), p. 339.
Harmon, H. McQuilkin, *The Evolution of Psychic Healing.* San Jose, California, 1909.
Hartmann, Franz, *Paracelsus.* London: Redway, 1887.
Harvey, E. Newton, *A History of Luminescence to 1900.* Philadelphia: American Philosophical Society, 1957.
*Helio-magnestat.* Mimeographed brochure of the Nicholson, Brown Company, of California.
Huard, Pierre, and Ming Wong, *Chinese Medicine.* New York: McGraw-Hill, 1968.
Huff, P., *Cycles in Your Life.* London: Victor Gollancz, 1965.
Inglis, B., *Fringe Medicine.* London: Faber & Faber, 1964.
Inyushin, V. M., "Biological Plasma of Human and Animal Organisms." Unpublished translation of article issued by Novosti Press Agency.
James, T. F., "Cancer and Your Emotions," *Red Book* Magazine, June, 1960.
Johnson, Raynor C., *The Imprisoned Splendor.* London: Hodder & Stoughton, 1953.
*Journal of Orgonomy* (Orgonomic Publications, Inc., P. O. Box 476, Ansonia Station, New York, N.Y. 10023).

Karagulla, Shafica, *Breakthrough to Creativity*. Los Angeles: De Vorss, 1967.

Katsenovitch, R., "L'hydroaeroinisation lors du traitement de la phase non-active rhumatisme." Milan: Carlo Erba Foundation, 1969.

Katz, Jack, and Gallagher, Thomas, and others, "Psychoendocrine Considerations in Cancer of the Breast," *Annals of the New York Academy of Sciences*, Vol. 164, Art. 2 (1969).

Keleman, Stanley, "Bio-energetic Concepts of Grounding," *Energy and Character*, September, 1970.

Kelley, Charles R., *A New Method of Weather Control*. Stamford, Connecticut: Interscience Research Institute, 1961.

———, *Primal Scream and Genital Character*. Santa Monica, California: Interscience Work Shop, 1971.

Kent, Caron, *The Puzzled Body*. London: Vision Press, 1967.

Kepler, Johannes, *Tertius Interveniens*.

Koestler, Arthur, *Arrow in the Blue*. London: Macmillan, 1952.

Kornblueth, Igho H., "Artificial Air Ionization in Physical Medicine, Preliminary Report," *American Journal of Physical Medicine*, Vol. 34, No. 6, December, 1955.

Krueger, A. P., "The Biological Effects of Gaseous Ions," *Bioclimatology, Biometeorology and Aeroionotherapy*. Milan: Carol Erba Foundation, 1969.

———, Hicks, W. W., and Beckett, J. C., eds., *Weather, Climate and the Living Organism*. Amsterdam: Elsevier, 1962.

———, Smith, R. F., and Ing Can Co, "The Action of Air Ions on Bacteria," *Journal of General Physiology*, Vol. 41, No. 2 (Nov. 20, 1957), pp. 359–81.

Laub, E., "Man as a Magnet," *Journal of British Society of Dowsers*, March, 1958.

Laurens, Henry, *Physiological Effects of Radiant Energy*. New York: Chemical Catalog Company, 1933.

Laws, C. A., and Holliday, E., "Organic Electronics—the First Breakthrough Since Antibiotics," *Energy and Character*, September 1972, pp. 20–23.

Licht, Sidney, *Therapeutic Electricity and Ultra Violet Radiation*. New Haven, Connecticut: Eliz. Licht, 1959.

Linhart, J. G., *Plasma Physics*. New York: Interscience Publishers, 1960.

Long, Max Freedman, *The Secret Science at Work*. Los Angeles: Huna Research Publications, 1953.

———, *The Secret Science behind Miracles*. Los Angeles, 1948.

Lowen, Alexander, *Pleasure*. New York: Coward-McCann, 1970.

Maby, J. Cecil, *The Physical Principles of Radiesthesia*. Birmingham, England, 1966.

———, and Franklin, T. Bedford, *The Physics of the Divining Rod*. London: G. Bell and Sons, 1939.

Mayling, H. A., "Structure and Significance of the Peripheral Extension of the Autonomic Nervous System, *Journal of Comparative Neurology*, Vol. 99, 1953.

Michelmore, Peter, *Einstein, Profile of the Man*. London: Fred Muller, 1963.

Mills, John, *Realities of Modern Science*. Quoted in *Electronic Medical Digest*, Autumn, 1950, pp. 14–15.

Moss, Louis, M.D., *Acupuncture and You*. New York: Citadel Press, 1966.

Moutin, Dr., "Animal Magnetism and Contemporary Science," *La Chronique Medical*, May 15, 1897.

Nameas, J., "Nature and Possible Causes of the Northeastern U.S. Drought during 1962–65," *Monthly Weather Review*, September, 1966.

Neill, A. S., and others, *Wilhelm Reich*. Nottingham, England: Ritter Press, 1958.

*Orgone Energy Bulletin* (quarterly), published by Orgone Institute Press, 799 Broadway, New York, N.Y. 10003.

Ostrander, Sheila, and Schroeder, Lynn, *Psychic Discoveries Behind the Iron Curtain*. Englewood Cliffs, New Jersey: Prentice-Hall, 1970.

Osty, Eugène, *Supernormal Aspects of Energy and Matter*. London: Society for Psychical Research, 1933.

Panchadosi, Swami, *The Human Aura*. Chicago: Advanced Thought Publishers, 1916.

Pauwels, Louis, and Bergier, Jacques, *Dawning of the Magicians*. New York: Avon Books, 1968.

Piccardi, Giorgio, *The Chemical Basis of Medical Climatology*. Springfield, Illinois: C. C. Thomas, 1962.

Pierrakos, John, "The Energy Field of Man," *Energy and Character*, Vol. I, No. 2 (May, 1970).

———, "The Energy Field of Plants and Crystals," *Energy and Character*, Vol. I, No. 3 (Summer, 1970).

Poinsot, M. C., *Complete Book of the Occult and Fortune Telling*. New York: Tudor Publishing, 1945.

Prasad, A., *Zinc Metabolism*. Springfield, Illinois: C. C. Thomas, 1966.

Price, Derek J. de Solla, *Science since Babylon*. New Haven, Connecticut: Yale University Press, 1961.

*Proceedings*, British Congress on Radiesthesia and Radionics, London, 1950.

*Proceedings*, International Conference sponsored by Skin and Cancer Hospital, Temple University, Health Science Center, and International Society of Biometeorology.

Puck, I., and Sagik, B., "Virus and Cell Interaction with Ion Exchangers," *Journal of Experimental Medicine*, 97:808–820, 1953.

*Radio Perception, Journal of the British Society of Dowsers*, London.

Rahn, Otto, *Invisible Radiations of Organisms*. Berlin, 1936.

Raknes, Ola, *Wilhelm Reich and Orgonomy*. New York: St. Martin's Press, 1970.

Ramacharada, Yogi, *Science of Breath*. Chicago: Yogi Publication Society, 1904.

Ramsay, E. Mary, *Christian Science and Its Discoverer*. Boston: Christian Science Publication Society, 1923.

Randall, James, *Elements of Biophysics*. Chicago: Year Book Medical Publishers, 1958.

Ratcliffe, J. A., ed., *Physics of the Upper Atmosphere*. New York and London: Academic Press, 1960.

Ravitz, L. J., "Application of the Electrodynamic Field Theory in Biology, Psychiatry, Medicine and Hypnosis," *American Journal of Clinical Hypnosis*, Vol. I, No. 4 (April, 1959).

———, "Bioelectric Correlates of Emotional States," *Connecticut State Medical Journal*, Vol. 16 (1952), p. 499.

———, "History, Measurement and Applicability of Periodic Changes in the Electromagnetic Field in Health and Disease," *Annals of the New York Academy of Sciences*, 1963, pp. 1144–1201.

Redfearn, J. W. T., "Bodily Experiences in Psychotherapy," *Energy and Character*, May, 1970.

Reich, Wilhelm, *The Cancer Biopathy*. New York: Orgone Institute Press, 1948.

———, *Character Analysis*. London: Vision Press, 1950.

———, *Cosmic Superimposition*. Rangeley, Maine: Orgone Institute Press, 1951.

———, "Discovery of the Orgone," *International Journal of Sex Economy and Orgone Research* (New York), Vol. 1 (1942).

———, *History of the Discovery of Life Energy—The Einstein Affair*. Rangeley, Maine: Orgone Institute Press, 1953.

———, *The Oranur Experiment*. New York: Orgone Energy Press, 1951.

———, "Orgonotic Pulsation," *International Journal of Sex Economy and Orgone Research*, Vol. 3, Nos. 2 and 3 (October, 1944).

———, *Selected Writings*. New York: Farrar, Straus and Cudahy, 1960.

Reichenbach, Karl von, *Abstract on Researches on Magnetism and Certain Allied Subjects Including a Supposed New Imponderable*, translated and abridged by William Gregory. London: Taylor & Walton, 1846.

———, *The Odic Force—Letters on Od and Magnetism*, translated by F. D. O'Byrne. London: University Books, 1968.

Rejdak, Zdenek, and Krbal, F., "From Mesmer to the Fifth Physical Interaction and Biological Plasma," *Journal of Paraphysics*, May, 1971.

Reznikoff, Marvin, "Psychological Factors in Breast Cancer: A Preliminary Study of Some Personality Trends in Cancer of the Breast," *Psychosomatic Medicine*, Vol. XVII, No. 2 (1955).

Rindge, Jeanne, "Are There Healing Hands?" *Response* (Rosary Hill College, Buffalo, N.Y.), Vol. 2, No. 2 (Spring, 1968), pp. 18–21.

Rocard, Yves, *Le Signal du sourcier.* Paris: Dunod, 1962.

Rochas d'Aiglun, Albert de, *Le Fluide des magnétiseurs.* Paris: Carré, 1891.

Ruch, Theodore C., and Fulton, John F., eds., *Medical Physiology and Biophysics.* Philadelphia: Saunders, 1960.

Rycroft, Charles, *Reich.* London: William Collins, 1971.

Salzman, Leon, *Developments in Psychoanalysis.* New York: Grune & Stratton, 1962.

Schaefer, Karl E., ed., *Man's Dependence on the Earthly Atmosphere.* New York: Macmillan, 1958.

Schliephake, E., "Therapeutic Value of High Frequency Current," *International Conference on Physical Medicine.* Amsterdam: Excerpta Medical Foundation, 1964.

Schmid, Alfred, *Biological Effects of Air Electricity.* Bern: Haupt, 1936.

Schwarzschild and Bierman, "Static Electricity," *Medical Physics,* Otto Glasser, ed. Chicago: Year Book Medical Publishers, 1948.

Shapiro, Marc, "Mesmer, Reich and the Living Process," *Creative Process,* Bulletin of the Interscience Institute, Vol. IV, No. 2 (June, 1965).

Sinaya, M. S., "The Influence of Unipolar Air Ions on the Electrophoretic Mobility of Erythrocytes," *International Journal of Biometeorology,* Vol. II, 1967.

Singer, Stanley, *The Nature of Ball Lightning.* New York: Plenum Press, 1971.

Sivananda, Swami, *The Science of Pranayama.* Rishikesh, India: Yoga Vedanta Forest Academy Press, 1962.

Smith, C. N., ed., *Medical Electronics.* London: Iliff and Son, 1960.

Smith, Grover, ed., *Letters of Aldous Huxley.* New York: Harper & Row, 1969.

Sollberger, J., *Biological Rhythm Research.* New York: American Elsevier, 1965.

Solomon, George F., "Emotions, Stress, the Central Nervous System and Immunity," *Annals of the New York Academy of Sciences,* Vol. 164, Art. 2, Conference on Cancer.

*Some Unrecognized Factors in Medicine.* London: Theosophical Research Centre and Publishing House, 1939.

Still, Alfred, *Borderlands of Science.* New York: Philosophical Library, 1950.

————, *Soul of Amber.* New York: Murray Hill Books, 1944.

Sudre, René, *Parapsychology.* New York: Citadel Press, 1960.

Szent-Györgyi, Albert, *Bioelectronics.* New York: Academic Press, 1968.

————, *Introduction to a Submolecular Biology.* New York: Academic Press, 1969.

Tchizhevsky, A. L., *Air Ionization: Its Role in the National Economy*. Moscow: 1960.

———, *International Conference on Therapeutic Radiation*. Amsterdam: Excerpta Medical Foundation, 1964.

Thompson, Clara, *Psychoanalysis: Evolution and Development*. New York: Hermitage House, 1950.

Tocquet, R., *Cycles et rythmes*. Paris: Dunod, 1951.

Tromp, S. W., *Medical Biometeorology*. New York: American Elsevier, 1963.

———, *Psychical Physics*. New York: American Elsevier, 1949.

———, *Survey of Human Biometeorology*. Geneva: Secretariat of the World Meteorological Organization, 1964.

Turner, James A., *The Chemical Feast*. New York: Grossman, 1970.

Urbach, F., *Biological Effects of UV Radiation*. London: Pergamon, 1969.

Weatherhead, Leslie, *Psychology, Religion and Healing*. London: Hodder and Stoughton, 1951.

Westlake, Aubrey, "Further Wanderings in the Radiesthetic Field," *Journal of the British Society of Dowsers*, December, 1951.

———, "Vix Medicatrix Naturae," *Proceedings of the Scientific and Technical Congress of Radionics and Radiesthesia*, London, May, 1950.

Wilbur, Sybil, *Life of Mary Baker Eddy*. Boston, 1907.

Woodard, Christopher, *A Doctor Heals by Faith*. London: Mowbray, 1954.

Worrall, Ambrose and Olga, *Explore Your Psychic World*. New York: Harper, 1970.

# INDEX